Billy's halo

Billy's halo

Love, science and my father's death

RUTH McKERNAN

Joseph Henry Press
Washington, D.C.

Joseph Henry Press • 500 Fifth Street, NW • Washington, DC 20001

The Joseph Henry Press, an imprint of the National Academies Press, was created with the goal of making books on science, technology, and health more widely available to professionals and the public. Joseph Henry was one of the founders of the National Academy of Sciences and a leader in early American science.

Library of Congress Cataloging-in-Publication Data

McKernan, Ruth.
 Billy's halo : love, science and my father's death / Ruth McKernan.
   p. cm.
 Includes bibliographical references and index.
 ISBN 0-309-10100-X (hardcover) 1. McKernan, Ruth.
 2. McKernan, Bill—Death. 3. Neuroscientists—United States—Biography.
 4. Neurosciences. I. Title.
 QP353.M39 2006
 612.8092—dc22

                    2006004445

Printed in the United States of America. Printed in partnership with Transworld Publishers.

For my father

. . . my father would pick me up and hold me high in the air. He dominated my life as long as he lived, and was the love of my life for many years after he died.

Eleanor Roosevelt

Science and everyday life cannot and should not be separated. Science, for me, gives a partial explanation of life. In so far as it goes, it is based on fact, experience and experiment.

Rosalind Franklin, in a letter to her father (1940)

Contents

Acknowledgments

FOR BOTH STRANDS OF THIS BOOK—THE SCIENTIFIC AND THE personal—I am grateful to a considerable number of people. To the many scientists whose presentations I've listened to, whose papers I've read and whose company I have enjoyed while filling my head with information I say a heartfelt thank-you—especially those of you who were unaware that your contributions would end up here. In particular, I acknowledge the input of Gary Aston-Jones, Colin Blakemore, Phil Cowan, Gerard Dawson, Deborah Dewar, Jonathan Flint, Fred (Rusty) Gage, David Goldman, Elisabeth Gould, Les Iversen, Barry Keverne, Stafford Lightman, Robert Marcus, Don Nicholson, Randall Reed, Trevor Robbins, Larry Squire, Geoffrey West, Lewis Wolpert, the late Jeffrey Gray and the late Max Perutz. I thank Trevor Robbins for reading and commenting on an early version of the book. I have drawn from popular science as well as from popular scientists. Among many excellent books, I heartily recommend *Nature via Nurture* by Matt Ridley, *The Faber Book of Science*, *What Is Time?* by G. J. Whitrow, *The Universe Next Door* by Marcus Chown and *Just Six Numbers* by Sir Martin Rees.

I started out as a scientist and have had to learn how to be a writer. Thankfully, many people have helped me. I acknowledge the support of Steve Connor, Susan Watts and Tom Wilkie, who worked at the *Independent* during my Royal Society Media Fellowship and introduced me to the art. I am grateful to Jessie Lendennie and family at Salmon Publishing, The Arvon Foundation and my fellow students on two courses, notably Tim Hayward, Karen Huckvale and Dean Parkin. I thank my tutors, Jon Ronson and in particular Justine Picardie, who gently, kindly pointed out my path through the woods. I appreciate the advice, empathy and commitment of Anne Engel and my editor Marianne Velmans and colleagues at Transworld Publishing.

I want to thank my father's friends, particularly John Addison, Carole Francis and Brian Woods, for their contributions. At the heart of this book, though, is my family, my father's family. I thank my uncle, Tom, for the stories of Billy's early life, and my brothers Stuart, Ian and Andrew too. Each of them could equally have written a 'son's tale' and their accounts would be different from mine, but no less significant. Our stories would pale besides our mother's, though. Catherine was in every sense Billy's complimentary strand, and is the understated strength of our family; always, always there for us. I have portrayed her lightly here for her story is not mine to tell and would fill a book by itself. Thank you is too small a phrase for you, Mum.

To my children, Francesca and Adam, I say please forgive me for not being at the beach, the playground or the football match, for squirrelling myself away for the best part of three years to write this book. And for you, my husband Gerry, every reader should know the magnitude of your gift. You willingly took on every role in our life—from domestic dervish to personal psychotherapist—so that I could put the pieces of this story together. Without you, without your love, *Billy's Halo* would be no more than a fuzzy, fading thought.

Permissions

Billy's
halo

Introduction

MY FATHER WAS MADE OF THE SAME COMPONENTS AND GOVERNED BY THE same rules as any other man. He was a transient mix of DNA and proteins, swimming in a multitude of metabolites. That's all he was; it's all any of us are. As a biochemist, I know. I have spent two decades studying the molecules we're made of. Yet to see *him* in this way is like staring at his ashes.

Science can't reduce love to a pattern of neuronal activity nor express grief as a mathematical formula. Even Albert Einstein, the most exacting scientist of them all, knew that 'imagination is more important than knowledge'. How we see life is more important than what we know.

I experience events emotionally as daughter, wife and mother, and yet the scientist in me can't help but try to understand and explain. My job, working for a pharmaceutical company, is to find new treatments for disorders of the brain. The subjects I know well—anxiety, mood disorders, the malleable nature of neurons—adopt new meaning when they become part of my family's life. Memory, consciousness, neurophysiology, my academic and professional interests, all find sharper focus when relevant to my father's condition. Sometimes I gain comfort from that knowledge; it removes fear of the unknown. At other times fear of the unknown simply becomes fear of the known and I am left, as we all are, with little more than hope.

Science is fascination and awe. It allows a view through a different window, beyond the restrictive scale of our world, and this brings comfort, for in science I have found ways to see my father as immortal.

He will always be a part of me—the very way the brain works guarantees that. As we learn, permanent changes occur in the brain; neuronal architecture is modified by experience. As they say in forensic medicine, 'every contact leaves a trace'. For me, for my father's friends and family, his legacy is not in the business he left to his sons, nor in the plaque listing Bill McKernan alongside other past captains of Eltham Warren Golf Club. It is written in our heads. Those of us who experienced his influence would be wired differently without him. He is braided into my very being.

To an evolutionary biologist, our role on Earth is to ensure the propagation of our genes. As a gene machine my father—or Billy to his immediate family—was a great success; his personal packets of DNA endure in abundance. My three brothers—Stuart, Ian and Andrew—and I each have 50 percent of his genes and our ten children have another 25 percent apiece. We are a pretty prolific family; procreation is one of our skills. By my calculations, Billy's genes have multiplied 475 percent. Richard Dawkins would be proud.

But numbers don't tell the whole story. As Andrew has got older his walk has become more like our father's, Ian has inherited the capacity to hold an audience in the palm of his hand, while Stuart is the only one of us whose academic ability is truly any match for Billy's. And me? It's hard to see oneself as others do but that large raised freckle above my left eyebrow and the way my upper gums show when I smile are strangely familiar.

Genetically, my father has longevity, but, let's face it, a gene is such a tiny thing. It takes less than one thousandth of a second to make a gene, little more than seconds to translate it into a protein, and the machinery that does so is less than a thousandth of a millimeter in size—that's how small a gene is. In the scale of time and space, even a few hundred thousand genes isn't much of a legacy. My father's lifetime was 65 years and 354 days. In moments of grief, minutes more were all I asked. But really I wanted him to live forever. To see an oak tree grow and die he'd have had to live a thousand years. To see biological evolution would take hundreds of thousands of years and to witness shifting continents, the formation of galaxies, we're talking light years; the language of space-time and cosmology rather than biology. Forever is a very long time.

Professor John Brown, the Astronomer Royal for Scotland, once told me that when we look up to the stars at night their light takes so long to reach us that we are looking back in time. It takes eight minutes for light from the Sun to reach Earth, he said, and the next closest star to the sun is Proxima Centauri, four light years away.

If you could look down on my family from Proxima Centauri, right now my six-year-old son, Adam, would be sitting in his high chair, playing with his cereal. My mother would be writing Christmas cards early for those overseas and my father would still be alive. An observer from Sirius, 8.6 light years away, would see us gathering for his big sixtieth-birthday bash in Captiva Island, Florida. From most of our universe, the millions of stars more than a hundred light years away, my father has yet to be born.

I take comfort in the knowledge that there will always be a part of the cosmos where my father is alive. Somewhere in our world, right now, he is hitting his first hole in one. From other distant places up in the ether he is just sitting behind his desk as he did, most days, planning the future of his business. I can no longer see him from here but as time passes, when my daughter, or even my daughter's daughter, no longer walks this earth, the halo of Billy's existence will still be spreading slowly out into the universe.

Science is just one way of looking at life. But it's my way and for a while it was my father's too—or, at least, it was a language we shared. It wasn't until he died that I realized how intimately science and emotion merge. Of course, they are anatomically integrated; the hypothalamus, the amygdala and a whole compendium of hormones control our feelings, relaying information out to the rest of the body (and vice versa). Science and emotion meet metaphorically too; intimate details of physiology have parallels in the world of human conduct. The way the brain is built, the way it develops and operates, has similarities with the way communities organize themselves and make decisions. The way a cell can die—by choice or accident— finds parallels in human behavior, too. If life imitates anything, it's science, not art. So, now, as I tell our story through a collection of scientific themes, like pearls strung together on the thread of my father's life, I know he won't mind. In fact, part of me expects that, wherever he is, he'll still be proud.

1

Memory

Ah, Memory—that strange deceiver!
Who can trust her? How believe her –
While she hoards with equal care
The poor and trivial, rich and rare;
Yet flings away, as wantonly,
Grave fact and loveliest fantasy?

From 'Memory' by Walter de la Mare

NOVEMBER 2001, AND A POSSE OF SCIENTISTS CAN BARELY WAIT FOR A seventy-three-year-old American to die. It's not that we're heartless or mercenary, we just want to see his brain. Close up, in the flesh. The organ's owner, discreetly known only by his initials, HM, has been subjected to more neurological and psychiatric tests than any other living person. I've ogled at his brain scans in learned journals. Others have probed and prodded his cerebral circuitry, and the investigations will not stop at death itself. His brain is unique. HM is unique. His tragic story is a science classic, for it sparked off decades of research into how and where memories are made.

As a child, not much older than my ten-year-old daughter, HM suffered intractable epilepsy. It started in two inch-long structures deep within his grey matter: his hippocampi. (The name derives from *hippokampos*, sea horse in Greek, and it perfectly describes their shape.) In 1953, at the age of twenty-seven, HM underwent experimental surgery in Hartford Hospital, Connecticut, where the surgeon William Beecher Scoville removed the structure from both sides of his brain. The good news was that the operation virtually cured his epilepsy, but the bad news was that it left him a complete and total amnesiac with no ability to form new memories. When Dr. Scoville and his colleague Dr. Brenda Milner published his case history they described for the first time 'the importance of the hippocampal region for normal memory function'.

Without that lump of cerebral putty, HM loses more than just memories. He barely notices the passage of time because he can't remember what happened a few minutes ago. Scientists who study him have to re-introduce themselves and re-explain the purpose of their experiments every day. His nurse is a stranger each morning and he doesn't remember any conversations from her previous visits. He scarcely knows his age, having no perception of the decades that have passed since his operation. Like the principal character in Oscar Wilde's novel *The Picture of Dorian Gray*, HM expects to see a young man when he looks in the mirror; he barely recognizes his own face, despite the familiarity of shaving it every day.

Studying HM tells us something about different types of memory. For example, we know that emotional associations can form without a hippocampus because, although he can't remember a single episode since his operation, HM knows that his mother is his mother and he retains a strong sense of right and wrong. From all I've read he seems a likeable man with a warm sense of humor despite his wide-reaching memory and associated cognitive problems. When one researcher asked him what he tried to remember he replied: 'I don't know 'cause I don't remember what I tried!'

Life must be very strange without a hippocampus, living only in the present. Information is always fresh—there can be no old news. Such has life been for the most studied man in neuroscience. The same joke will be just as funny after the hundredth telling. Equally, the shock of bad news will evoke the same emotional reaction each time; the pain of bereavement will strike anew, not dwindle.

The pain of bereavement? Where did those words come from?

Scientists don't write about emotional things like the pain of bereavement. We write about facts, about biological knowledge: the hippocampus, the keyboard of the brain; the conduit through which all things new and exciting transit into documented personal history. We write about how the brain develops, how emotional and factual memory systems vary and why early experiences are hard to store without a well-developed language. But it is the personal, the way science affects us, that brings the subject to life.

'Who's Granddad Bill?' asks Adam.

The question comes from nowhere while I'm tidying up the bedroom. It's a small, simple question, the type a six-year-old asks with little thought at all.

'When I go to church with Daddy we always light a candle for Granddad Bill.'

Like any developing child, my son needs a limitless supply of facts; more input to mold his brain. Content is inconsequential, volume is all.

'You remember Granddad Bill?' I say.

He innocently shakes his head.

'He was my daddy, don't you remember? I know you were only three when he died but you *must* remember him.' I want to shake him by the shoulders until the memories fall into place. Instead, I pair the socks in my hand and put them away in the top drawer. Three years? Has it really been three years since my father died?

Adam looks at me quizzically, straining, trying to remember.

I take a picture off the dressing table and poke my finger repeatedly at the glass. 'That's Granddad Bill, sitting on the chair with Francesca. You remember when we went to Florida. You remember Disney World, don't you? Surely you remember Granddad Bill!'

His eyebrows furrow and he stares at me, displaying no hint of recollection.

Moments later he almost whispers: 'Did Granddad Bill like me?'

Irritation turns to empathy. I want to kiss him. There is something there. Somewhere in that beautiful, but half-baked, brain is some element of a memory of my father.

Billy hadn't found it easy to relate to small children as he grew older. He wasn't the cuddly figure to his ten grandchildren that he had been to me at their age. He used to cavort around the living room with me, giggling, on his back, toss my convulsed torso onto the sofa and then rub his Saturday morning stubble across the gap in my pajamas until I shrieked for mercy. Some three decades spent running the family business and speaking with executive-style gravitas only to adults robbed him of that playfulness. There was little space left for entertaining toddlers; he'd sort of lost the knack. Even the more relaxed persona that he presented on holiday didn't quite hang together: a granddad who pretends to scare you by roaring like a tiger is only funny when you know he's pretending. My father was a bit too convincing and Adam was a bit too insecure for that.

'Of course he did,' I said, pulling him toward me. 'He loved you very much.'

Vocabulary and memory go hand in hand. Studying HM shows us that the hippocampus is necessary for both. My son recalls something of the emotions his grandfather evoked in him, but with only the rudiments of vocabulary and grammar he couldn't have formed the same sharp quality of memory that you and I would. In the three years since my father died, Adam's verbal repertoire has continued to grow to the point that a little respite from the constant chatter would be a pleasant change. HM, on the other hand, has learned no new words since his operation over fifty years ago. Indeed, his vocabulary can't expand. Consequently, he doesn't have a single personal anecdote in his head.

Another unfortunate soul, known as EP, whose hippocampus was destroyed by encephalitis, also has no idea who people are, no sense of time and no understanding of where he lives. But

EP's lesion is less extensive than HM's and he can remember information prior to his illness. He remembers the town where he grew up every bit as well as his boyhood friends. EP shows us that while the hippocampus is absolutely necessary for forming memories they aren't stored there. You don't need a keyboard to read back what you've typed.

Memories of place and time, and vocabulary to express them, all need an intact hippocampus; a connection that the poet Seamus Heaney might have been making when he described the early experiences important in his work. 'Memory is where space and time and language come together,' he said. There is neurological truth in the poet's words. Episodic memory comprises the 'where, when and what' of an event. The hippocampus is where these three meet—for the 'sea horse' in the brain is essential to their formation.

If only I could remember more about my father; more about the times we spent together, the way he looked, the sound of his voice and the feel of that Saturday morning stubble.

There are ways to commit things deliberately to memory. I know, at least in theory, how to try and emulate the legendary Shereshevskii and his apparently inexhaustible memory. The Russian genius could remember limitless numbers, words or even nonsense syllables. He could study a fifty-figure table for just three minutes and then reproduce it exactly in less than forty seconds.

There are some rare individuals gifted with a hugely superior memory—Shereshevskii may have been one. Most of us, though, can still learn to improve our memories. Eleanor Maguire, a neuroscientist from the University of London, scientifically debunked the notion that some people possess a special photographic memory. Her imaging studies, which tested the brains of eight top competitors in the World Memory Championships, showed that their brains are structurally no different from those of the rest of us and their memory is improved by training. The trick is to link every word to a graphic image, distribute them along an imaginary path and recall them by retracing your steps later. Dr. Maguire showed that this mnemonic technique uses more regions of the brain than other types of recall and these are the regions involved in spatial navigation, which makes perfect sense.

The 'mental walk' technique itself isn't new. The Greek poet Simonides of Ceos reportedly used it as long ago as 500 BC. More recently, Andi Bell, the 1998 World Memory champion, used this and other techniques to memorize a whole deck of cards in 34.03 seconds.

If I wanted to remember a string of facts, whether it be the chronology of kings or a shopping list, the mental walk technique would be just right—but that's not what I want to remember. I want to remember more of the important little details, like the way the creases around my father's eyes deepened when he smiled, the way his left eyebrow was a bit bushier than the right, with a few extra-long hairs sticking up in the middle as if he were always just a little bit surprised. I want to remember the way he looked when his thinning grey hair was newly cut and when exactly it was that my brothers started calling him 'Baldy Billy', a taunt they discreetly shortened to 'Billy' for public acceptability down at the rugby club, where even the vice-presidents need a nickname.

I know enough to know that developing my memory skills won't help me now. There are many tricks for remembering new strings of facts but they won't help me pull out existing memories of the past. Nor do they have the capacity to illuminate even the smallest scene with the sensations and emotions that accompanied it.

I hear the telephone ringing. Telephones don't ring at seven on a Saturday morning, not unless it's trouble. My husband, Gerry, answers it and, before he says a word, I know. I can physically feel it, like the wind rushing through a station before the train arrives.

It will be about my father. It will be either a stroke or a heart attack. I weigh up which is more likely: the head or the chest? An incontinent future spent dribbling soup from the confines of a wheel-chair or major surgery followed by a slow uphill struggle until the next collapse? The clinical details matter little for I give my father short odds of surviving either.

I reach out for the receiver, seeing my father as I last saw him: a behemoth of a man, over six feet tall and slightly hunched from years spent bowed in devotion to his two passions, work and golf. His ample frame appears in the hallway of his home, dressed in a powder-blue Pringle sweater and black-and-white checked slacks. He has been for a quick nine holes in the late afternoon at his local golf club. He has more time for golf these days since his supposed retire-ment—although real 'slippers and pipe' retirement will always be a stranger to my father. Men like him don't hand over the family busi-ness to their sons and walk away. They hover and want to be needed. They make themselves busy and find windows of opportunity to talk to other very busy people.

He is wiggling his fingers in the characteristic way that he does; more movement than a tremble, as if he were mentally rehears-ing a performance of Beethoven's 'Für Elise'. He isn't aware that his fingers move when he thinks, in the same way that a child doesn't sense his lips are moving when he reads to himself. Billy's fingers have a few wiry hairs between the joints and marginally too long, calcareous, tough old fingernails. Is the flesh slightly puffy, perhaps a little blue? That would be a sign of heart problems, wouldn't it?

He stoops down, half sitting on the kitchen chair, to remove his size 12, casual but smart Italian loafers. He puts on his slippers—maroon leather with real hide soles—the type worn by the alpha male who takes off his tie, casually replaces it with a paisley cravat, and slips into something a little more comfortable after a hard day at the office, dear.

My father settles down to the last few clues of the *Daily Tele-graph* crossword that eluded him that morning, holding the folded newspaper sufficiently far in front of him to confirm that the lenses in his reading glasses need to be stronger. As he thinks, he absent-mindedly scratches at the skin under his left eye. He has a lot of skin under his eyes. Where other people have bags, he has hold-alls. If he had a mustache he would have fiddled with that. If he had more hair he might have run his fingers through it. Instead he scratches at the folds of skin hanging over his cheekbones. He has some funny little habits. In anyone else they would be annoying, but in my father they are part of his charm.

My mother's distant voice on the telephone tells me he's in Addenbrooke's Hospital. It started yesterday evening as a bit of a temperature, she says. Billy thought he might be coming down with the flu so he had a bath, took a couple of acetaminophen and went to bed.

'When I came in from babysitting he was very hot. His face was all swollen. He couldn't settle; he kept getting up, going to the toilet, then lying down again. He was sick, had diarrhea. I phoned the doctor. He said to call an ambulance just to be on the safe side. Ruth, you should have seen him. His face was so swollen on one side he couldn't speak properly. He wasn't making any sense.'

Swollen face? Mental confusion? Could be a stroke? Doesn't sound like a heart attack. But vomiting and diarrhea—they don't seem to go with either condition.

'The doctor was very good, though. Not our usual doctor. He was from Gold Street. Dr. Lort—do you know him? Young chap, very nice. He gave him a couple of injections before we got in the ambulance. Your dad wasn't making any sense, barely knew where he was. They're keeping him in for observation. The doctor said it could be an allergic reaction or an infection. I don't know how long he'll be in there.'

My first instincts must have been wrong. No sudden chest pain, no numbness. Not the right prognosis. We scientists are supposed to be good at analysis. We're supposed to try to pull information together from different places. We're taught to think rationally, logically, even laterally, out-of-the-box. But that's to answer questions of biology, not to recognize or treat medical conditions; I have a Ph.D. in molecular pharmacology, not medicine. All I can do is think. So I think. I think desperately hard, trying to dredge up anything useful. Allergy—a distant dog-eared, red ring-binder of college notes; the words I learned by rote for first-year physiology and barely understood. Infection—a relatively inconsequential ailment, hardly life-threatening. Penicillin works if it's gram-positive. Stroke—still worth pursuing all avenues until a line of research convincingly comes to an end. Ten years ago, I carried out some research as part of a team developing a drug for stroke. The project failed when we discovered that MK-801, as we knew it, while looking very good in reducing brain damage, unfortunately also had other effects that we couldn't explain. I had sat through so many lectures on stroke but it was all theory and so long ago. Perhaps it would be different if I'd met someone who'd actually had one. No, there's nothing there to help.

I tidy the lighthouse-shaped stencil and the little pots of blue and white paint into a neat pile on the kitchen table. Francesca and I will have to postpone our plans to decorate Adam's bedroom this weekend.

'I'm going round to see Grandma,' I say. 'Granddad's not very well.'

'Then you should go and see him too?' she replies in a manner that even at the time I remember thinking was somewhat precocious for a seven-year-old, but not entirely out of character.

'We probably will. He's in the hospital,' I reply in my most matter-of-fact voice, trying not to alarm her.

My daughter picks up her favorite lambswool glove puppet, which she affectionately calls Lamb Chop, and pops behind the cardboard Punch-and-Judy set. 'Can we come?' she asks brightly, swinging the grubby-faced mitten wildly from side to side.

'Maybe next time.'

The lamb's head disappears under stage level to be replaced by its auburn-haired controller. 'What's wrong with Granddad?' she says, almost as an afterthought.

'Well, we don't know yet,' I say, masking my deep concern. 'I expect we'll find out soon.'

She skips into the playroom, drops Lamb Chop into a box of equally unlikely characters, sits down at her own, child-sized table and starts drawing. Oh, the joys of a youthfully short attention span! I pick up her flute and put it back in its case, fold the surplus newspapers and grab my car keys.

'Give me a call as soon as you know anything,' says Gerry, hugging me fleetingly on the way out.

I remember that day vividly. It was three years ago and the memory is still super-real. Everything about it is magnified. What were small details have become significant and important. What was important has become too big to see in focus, too strong to bear.

Memory is not one single phenomenon, as I found out when I spoke to Larry Squire, the eminent Professor of Psychiatry from the University of San Diego. He explained that, in broad terms, there are two distinct types of memory (although for real experts these can be further subdivided into multiple categories). The first is known as declarative, or episodic, memory. It's what we think of intuitively as memory. We use it for facts and figures, for the particulars of past events of which we are consciously aware. It would be fine for remembering that my father's birthday was February 8, that he was admitted to hospital on Saturday, April 14, 1998, and that he was over seventeen stone at the time. Declarative memory was described by Oscar Wilde in *The Importance of Being Earnest* as 'the diary that we all carry about with us'. It contains details of emotions, thoughts and analysis in addition to bare facts. Declarative memory allows me to describe the events of that Saturday morning in detail and to remember all the associated thoughts and sensations.

I first started to learn about memory at the American Society for Neuroscience in November 2000. It was not a great introduction to the subject. The Miami Beach convention center was bigger than my home town. It could have swallowed Saffron Walden's town hall, the market square and the library and still have had the multistory car park for dessert, so by the time I found room A209 the session entitled 'Cognition: human learning and memory' was already well under way. I should really have been at a parallel session on the structure of ligand-gated ion-channels—my own area of expertise.

Instead, recent events drew me to learn more about memory. Each of the eleven fifteen-minute talks comprised a jumble of words, all of which I recognized but couldn't understand when assembled into such odd phrases as 'the remember-know recognition paradigm' and 'semantic consolidation precedes retrieval'. The shorthand phrases used by one set of experts seemed like little more than random words pulled from a textbook and linked together without the benefit of anything as common as a preposition or a vowel. Who would have thought that biochemistry and psychology were so far apart? I was only a few steps out of my own territory and the land was as foreign to me as particle physics.

Larry Squire was the chairman of that symposium and I learned much more from talking directly to him. The second type of memory, he explained, is called procedural memory. It is much less obvious and you might not even realize that it is a form of memory. We use it for storing how-to-do things. It isn't going to help me remember anything about my father but it will retain information on how to swim, ride a bike or solve a puzzle. It's virtually subconscious; the kind of memory that Walter de la Mare appreciated when he wrote that 'her exquisite vigilance enables me to walk, sing, dance'. Procedural memory is used for our habits, preferences or phobias. This kind of memory underpinned my father's ability to play golf, and its continual refinement got him to a handicap of eight at his peak. It was also reflected in his appreciation of a glass of Springbank malt, diluted one-to-one with water, no ice.

Different types of memory use different parts of the brain. We know that declarative memories require at least a functional hippo-campus—the case of HM showed us that. He also showed us that it isn't needed for procedural memory. When he lost his memory, HM didn't forget how to walk, sing or dance and he could still learn new actions, like tracing the outline of an object by looking in a mirror, for example. This is not an easy task and it requires repeated prac-tice—try it yourself and see. Once he had acquired the skill, though, HM didn't forget it and could still draw back-to-front a year later with no problem. The memory was stored away somewhere, and that somewhere didn't include the hippocampus.

There is no single part of the brain that contains all our memories. They aren't stored in a neat filing system; categorization is neither alphabetical nor thematic. Memories are encoded in the part of the brain that is most relevant for the task. So the memory for how to swing a golf club is stored in the cerebellum, the apple-sized chunk at the back of the brain responsible for the movement of our limbs, while memories of events are more likely stored in the cortex, at the front of the brain, which is involved in analyzing and processing thoughts. That's where I can find the handmade blue tiles of the Spanish restaurant just off London's Regent Street where my parents took me for my first grown-up dinner, the bath-sized dish of paella on the red checked tablecloth and the teenage embarrassment of my father asking the roving Mariachi band to sing 'Guantanamera' while I stared at my new brown leather birthday boots.

I might hope that all the memories of my father are there, some-where, in my frontal cortex just waiting for the right trigger to be released. For this to be true, it would mean that we would have to remember everything and be able to recall at will only what we wanted or needed. It would mean that we have the kind of memory that Morecambe and Wise joked about in one of their sketches:

ERIC MORECAMBE: 'My wife has a very bad memory.'
ERNIE WISE: 'What does she remember?'
ERIC MORECAMBE: 'Everything.'

It is a nice idea but, really, what would our minds be like if we collected every fact, event and feeling we ever encountered? Think of the level of clutter we would accumulate. Ultimately our minds would resemble a kitchen so cluttered that the surfaces were piled high with utensils, papers, objects, gadgets and dirty dishes; a functionless room where there was nowhere to put anything down and certainly no space to cook or eat. Our brains could not handle that level of detritus. Slowly, inevitably, we would cease to function.

Clogging our brains with unnecessary repetitive information could grind us to a halt just as easily as the world's computers did the day the Love-Bug virus was unleashed. How would we function if we couldn't separate the important from the endlessly trivial? What help would it be for me to recall the minutiae of each time I brushed my hair or drank a cup of tea? Having a totally inclusive memory didn't help Funes the Memorious, a fictional character created by Jorge Luis Borges. He could remember the exact shape of a cloud and compare it with the detail of the marbling on a leather-bound book that he came across weeks later. He missed nothing. Ultimately, though, Funes's prodigious memory destroyed him. The weight of his cerebral garbage drove him to an early death.

It may not be optimal to remember everything, so how about if we could *choose* what to remember? Imagine what life would be like if we had complete control over our memories—if deciding what to lay down for longevity were as simple as filing papers. Just think 'in' to keep it and 'out' to throw it in the mental wastepaper basket. It could be the end of shopping lists, lost car keys and knotted hankies. Examination revision could be done in one swift read-through of the course notes. Whole categories of excuses would be redundant. 'I forgot my homework' or 'I lost your phone number' would be no more.

Think of the benefits of deleting things we'd rather not recall, like the tuneless repetition of Elvis Presley singing 'Always on my mind', which has unfortunately been always on my mind since I last visited my old friend Debbie and the venerable Dr. Dewar demonstrated her new UX-D88 micro-component system's ability to belt out 'the King's' latest CD remix. Better still, I could dispose of that embarrassing and ill-advised liaison at the International Pharmacology Congress in 1989, when I was an innocent, unattached Ph.D. student. Wouldn't it be great if I could just think 'out' and have the whole sorry episode removed forever?

The ability to pick and choose what to remember does not exist. The brain just doesn't work like that. Mostly, what sticks is involuntary, unconscious and outside our control. No, what is written in our personal histories is not of our choosing.

My mother sits in the kitchen, in the same chair where I saw my father put on his slippers the day before. Her skin is a sleepless grey. Her legs are crossed at the ankles and tucked neatly under the chair with one hand slotted defensively between her knees. She takes a mouthful of cold coffee and releases another burst of loosely connected facts and opinions.

'Two doctors were waiting on the pavement outside Emergency Admissions when we arrived in the ambulance. We didn't have to wait at all. The man who did the brain scan was very nice. But your dad didn't understand what he was saying.' She takes another mouthful from the heavy stoneware mug. 'He hasn't got any pajamas, you know—not a single pair. He has never worn them, never been in the hospital before, either. We'll have to wait until Gray Palmer opens and get some on the way in.'

I nod. It wouldn't have been my plan—I'd rather go straight to the hospital and see my father for myself—but I say nothing, under the circumstances. My mother needs those pajamas. They are a symbol of hope; maybe he'll be sitting up in bed, drinking a cup of some standard-issue refreshment, by the time we get there—and maybe he'll be well enough to care what he's wearing.

She lights another cigarette, looks at me apologetically and empties the overflowing ashtray.

Some events are obviously very important, while others are inconsequential, yet it might seem that we remember both equally. Why would I remember the dated mannequins in Gray Palmer's window, the assistant's half-rimmed glasses hanging from his neck by a black cord or the azure, polyester-cotton pajamas, size XL, with navy piping around the collar? Why should I remember the shiny one-penny piece the assistant put into my hand and the thoughts I would never say aloud, far less sell?

Why is it that we remember some things better than others? Again, I found the answer in the words of Professor Squire. He explained that three factors encourage stronger memories to be made. These are repetition, attention and emotion.

We all remember nursery rhymes from our childhood because we have repeated them so often. So it comes as no surprise that repetition improves memory. How many times did I reach up for my father to kiss me on the cheek? Not that many when I was a child— he preferred the top of my head—but on my reaching adulthood he developed the habit of kissing me on the cheek every time we parted. When I say he kissed me on the cheek, I mean he actually kissed me on the cheek—lips to skin, not cheek-to-cheek and lips-to-air; close enough for me to appreciate how the smell of Aramis aftershave changes when blended with the solvent-extractable chemicals in his skin. Let's say that I saw him, on average, once a week. Not counting the ten years I lived some distance away, when studying in London or working in California, we exchanged kisses at least fifty times a year for eleven years; his lips touched my cheek five hundred times.

Repeated learning can expand the part of the brain used for the task, as demonstrated by the superiority of the London taxi driver's brain. Highly trained cabbies have an organ to brag about. Acquiring an intimate understanding of London's streets, buildings and traffic systems is known as 'doing the knowledge': repetitive, spatial learning on a grand scale. Forming memories of places and maps can be narrowed further to one particular part of the hippocampus. Dr. Eleanor Maguire and her colleagues imaged the brains of a group of taxi drivers in the same way they studied the memory maestros. They found that one subdivision of the hippocampus is bigger in qualified taxi drivers than in other men of similar age and background.

Whatever the reason, it's the kind of observation that can't help but make you think. Perhaps those with good spatial skills are drawn to be taxi drivers. Perhaps Fred Housego, the London taxi driver famous for winning *Mastermind*, had an advantage over other contestants because his job improved his memory. Or, perhaps, intensive repetitive learning actually encourages the growth of new brain cells, or expands those that already exist. The auditory cortex is larger in blind people. The motor cortex, involved in delicate hand movements, is larger in concert-level pianists and violinists. And maybe there's a tiny pixel in some distant corner of my brain that's permanently swollen by my father's frequent kisses.

Repetition, attention and emotion: these are the three basic rules for forming stronger memories. But how do neuroscientists know that? How do they measure memory in people—or animals, for that matter? How did they work out what affects memory?

In the case of spatial memory—how the taxi driver finds his way around London or the rat finds his burrow after a hard day's foraging—there are some important differences between animals and humans. Animals rely more heavily on smell than we do, for example. But in other respects we find our way around by using the same techniques; by using landmarks as cues to guide us.

The most widely used test of memory in rats is the Morris Water Maze, named after its inventor Professor Richard Morris from the University of Edinburgh. It works by putting rats in a high-sided circular pool, about two meters in diameter and filled with opaque fluid. The rats swim around until they find a platform, submerged just below the surface, on which to rest. After five to ten trials (starting from a different location each time) rats remember where the platform is and swim straight to it. They learn its position relative to cues on the wall around the pool in much the same way that sailors used to work out their ship's position by triangulating on the stars, or London taxi drivers know that if they have the River Thames on their left and the Savoy Hotel on their right they are heading west toward the Houses of Parliament.

The water maze is no longer just for rats. Scientists at London's Institute of Psychiatry have managed to convert the test into a form fit for *Homo sapiens* and my husband was one of the first to try it. He didn't actually have to swim; he had only to find a particular place in a virtual-reality maze. The maze is shaped like a bullring with abstract patterns on the wall. At first he was confused and disoriented, but after ninety minutes of training he could find the designated spot in less time than it would take him to get through level one of *Super Mario*. The signal from the fMRI scanner, synchronously measuring his brain activity, clearly showed strong hippocampal activity while he was learning. These are the kinds of tests neuroscientists use to understand which brain regions form memories and how they are stored, consolidated and recovered.

Nine thirty a.m. and I am scrabbling around in my handbag, looking for change to buy a parking ticket. My mother may have a fifty-pence piece in her purse but she has already passed the bicycle racks and parking bays for the disabled and is disappearing through the hospital's canopied entrance. The stride of her size-four Bally flats lengthens, almost breaking into a run—a motion that her legs haven't performed in years but, reassuringly, can just about manage.

I catch up with her in the corridors marked by artwork and follow closely on her shoulder. On the left corner, we pass a picture of a crouching young girl, squashed into its frame, as though the artist meant to put the bright pastels on a bigger piece of paper. We pass the dour, mud-colored picture of a Second World War plane and the phone booths on the right with their walls offering a plethora of leaflets: '*The Samaritans can help you*', '*Addenbrooke's Chapel—*

times of service' and smaller cards announcing: Dolphin Taxis, A1 Cars, chiropody at home. We stop at the lift, go up three floors, turn left toward the poster of two smiling nurses and through a pair of heavy double doors. We know where we are.

My pulse is racing. I clutch the Gray Palmer carrier bag. I prepare to smile. The words 'Hi, Dad, how are you?' are halfway to my lips. We step past the screen surrounding my father's bed. Object—lens—retina—visual cortex—thalamus—and on out to all other regions. In the fraction of a second it takes to process the scene in front of me I abort the words and bitterly regret buying the pajamas. I know instantly that the term 'kept in overnight for observation' in no way conveys the gravity of my father's condition. To say he is unwell would be like saying small-cell lung carcinoma is a bit of a chest complaint. An overnight stay in the hospital is never going to fix it.

Retelling a memory depends on language. Without it, nuance, detail and tone are compromised. Consequently memory, in people, is most easily explored using language. It is frequently tested by remembering lists of words: related words, different words, emotional words, nonsense words. But the relationship between language and memory is a complex one. If we lose our language we do not lose our memories. Instead we might inhabit a world as bizarre as HM's but perhaps even more frustrating. When Sir John Hale's stroke removed his ability to speak, it left his memory intact. The art historian still knew his wife, his friends and his books; his wealth of knowledge was unchanged. Even though he couldn't remember the word for 'sock' he could draw a diagram of how to get to the sock counter in the local department store, illustrating that spatial and verbal memory are distinct.

Common memory tests, such as recalling lists of words, helped to prove Professor Squire's second requirement for forming stronger memories: attention. In 1933, Hedwig Von Restorff showed that we remember things that stand out as different, unanticipated. When asked to remember a list of, say, fifteen words, people will remember those at the beginning and the end better than those in the middle, unless there is something unusual about one of the words that attracts attention. In the list Porsche, Dodge, Fiat, Ferrari, BMW, Mazda, Jeep, Daffodil, Citroën, Saab, Chrysler, Honda, Subaru, Ford, Mitsubishi, daffodil stands out because it is the odd one out. We pay attention to the unexpected.

My father's face looks different, as though it has been sculpted in unbaked dough. I feel the urge to press his cheek to test whether it's fully cooked. Any distinction of normal features in that huge puffy mass is long gone; his eyes can barely open because of the swelling around his neck and cheeks. His mouth won't close and his gums, normally only seen when he smiles, are permanently on display. He has more in common with the elephant man's less attractive older brother than with the man I know.

He doesn't greet me. He doesn't even seem aware that we have arrived; we are part of a parallel world that exists around him but doesn't reach him at all. Nothing outside his own body seems able to claim my father's attention. Our words don't reach his ears. My mother gently strokes his arm and he shows no reaction. He tosses and turns on his bed, sits up, lies down again and then gets up and walks to the end of the ward, no more than ten steps. It is not his normal walk. His toes usually point slightly inward and his step has rhythm. This is a shuffle, a bear-like lumber, his flat feet barely lifting from the ground between steps. But his gait is even: he isn't favoring one side over the other. The likelihood of a stroke wanes. Where the ties on his regulation hospital gown fail to meet, his exposed back reveals a stormy sunset of skin, patches of bruised blue merging into hot crimson. God, why does it look like that?

I ask the nursing staff about his condition during the night. They reply with sympathy instead of facts. A neighboring patient tells me that he wrenched his intravenous line out so many times that the nurses gave up trying to put it back in again, that he spent the night wandering up and down the ward in varying states of frenzy, tugging at the sleeves and neck of his flimsy hospital gown, ripping it and finally pulling it off.

'God, if he knew what he was doing he would die of embarrassment,' my mother says without a trace of irony in her voice.

'I'm sure he won't be able to remember any of it,' I reply with as much confidence as I can muster. His dignity was left at the front door like a wet umbrella.

My mother touches his arm lightly and is able to guide him back to the allocated bed, apologizing to the patients on either side as they pass. The poet Roger McGough could have been right here with me when he wrote: '*Hospitals are packed with it, saw my mother wracked with it. Pain—I hate that stuff.*'

I feel the urge to apologize too. I want to tell the gentleman in the opposite bed that my father is a kind and thoughtful man—a founder member of the golf section of the Duke of Edinburgh's award scheme, I'll have you know—a man who would be horrified by his own behavior. The nameless patient, who is at least wearing his own pajamas, nervously fingers the oxygen mask in his lap as if it were a cocked gun at the ready, insurance against some anticipated onslaught. On his face a knowing look says no apology is necessary. In the land of the sick, blame has no place.

'Dad, would you like a drink of water?'

He nods and feels for the plastic tumbler with both puffy, outstretched hands. Relieved that, this time, he can hear me, I place it shakily between his palms and watch it glide clumsily to his lips. He exhales a primitive, appreciative grunt, aware of the comfort a mouthful of cold water brings. He is also aware that he has dribbled down the side of his bulbous chin and struggles to catch the drip with the edge of his cup. He still knows what manners are and has the will to preserve them.

Attention matters. As I sit at my father's bedside, vigilant, on red alert, it is as if I have been frivolous with my life up until now. It is time to be serious, to grow up; everything else has been a dress rehearsal. I remember the words of a friend, an underwater diver, who said: 'Panic kills.' Maybe the opposite is true. Calm saves? I do my best to appear calm, though every nerve in my body is screaming for help. If I can stay calm maybe I can think, act, do whatever is necessary to help him, bereft as I am of any idea what that might be.

Despite intensely scrutinizing my father's every move, it isn't vice-like attention that strengthens this memory. It isn't the extraordinary facial features, his uncharacteristic walk or the state of his back, although these are undoubtedly important. No, it is Professor Squire's third proposition—emotion—that explains why I remember the events of April 14, 1998, in such detail.

Emotion is the powerful force that drives our behavior, our reactions—and, yes, even our memories. Its ability to enhance memory is less obvious than repetition or attention because it isn't a tool we consciously use. We can repeat irregular verbs and pay close attention to the teacher's pronunciation but we can't deliberately increase our emotional state in order to improve our French. Nonetheless, emotion is incredibly important. That's why we remember where we were when Kennedy was shot, when Diana lost her life or when we heard about 9/11. The huge emotional content of such events marks them strongly in our minds. Emotion is the tool that writes our mental diary. Indifference barely marks the page. The unusual might warrant a soft HB pencil. Happiness or unhappiness would take a darker ink, while fear, love or desperation stand out, highlighted in a fluorescent yellow marker. The more emotional an event the more firmly it is written.

The American neuroscientist Joseph LeDoux spent a decade tracking down the part of the brain responsible for such power. The 'emotional center' is a nut-sized bundle of cells, two inches in from the ear, called the amygdala (from the Latin for almond). His research group's initial experiments connected one particular emotion—fear—with memory. Rats were taught that when a light was switched on in their cage it would soon be followed by a mild electric shock to the grid under their paws. They soon learned that the light predicted an imminent shock and they quickly moved to avoid it. Without an intact amygdala, rats were unable to remember the pairing of light and shock and so stayed put. While they could remember plenty of other things, and could find their way around a Morris Water Maze without difficulty, they didn't show fear-related learning. Since then Professor LeDoux's team, and others, have found that the link isn't just causal, it's physical. There is a well-honed neural circuit, a direct line, connecting the amygdala to the hippocampus.

We have evolved a system that automatically remembers important things better. For so many aspects of science we look to evolution to understand why things are as they are. Evolutionary pressure is one of those rarely transgressed principles. If there could be an advantage in terms of survival, then scientists find a theory plausible. The effect of emotion on memory seems to fit the bill. High levels of stress hormones (adrenaline, glucocorticoids and others—of which more later) signal stronger memories to be laid down. Stronger memories of threatening or arousing events could make man more likely to survive and hence to reproduce. It would be just as useful for *Homo erectus* to remember which cave was home to a man-eating lion as it would be for twentieth-century man to remember the face of the armed attacker in the street.

Scientists and writers alike agree that knowing what's worth remembering and what isn't is the all-important choice. This was obvious to Descartes, who wrote that 'the usefulness of all the passions consists in their strengthening and prolonging in the soul thoughts which are good for it to conserve'. Passions to Descartes were what emotions are to us today. Three hundred and fifty years ago, long before we knew anything about the amygdala or the chemical nature of hormones, the French philosopher recognized that the value of emotions was in helping us to remember important things better.

More recently the Harvard professor E. O. Wilson described

emotion in more scientific language as 'the modification of neural activity that animates and focuses mental attention'—a starkly impassive definition of our human feelings. Nevertheless, whatever form of words you use, it's clear that our brains are programmed to remember traumatic or highly charged events and there is little we can do about it. We do not will ourselves to remember emotional things better, nor can we stop those memories being made. Once they exist, it is as hard to remove them as it is to unlearn riding a bike.

Each period of time I spend kneeling at my father's bedside stretches in length. My mother and I take it in shifts—fifteen minutes at his side, then five to ten in the peace of the relatives' room for some respite. Occasionally she goes outside for a cigarette, which she invariably stubs out to hurry back, just in case something happens. Nothing much does. My father's actions become less frenetic. Time slows as he does. I ask him endless, pointless, useless questions, just to see if he can understand me. His grunts become sighs; the sighs become silence. He is oblivious to our presence and doesn't respond even when the nurse speaks in the slightly too loud, over-emphatic manner that one usually reserves for chastizing small children or giving directions to foreign tourists.

'Please, please, call the doctor. He isn't getting any better.'

At last the nurse picks up the phone again. I can't make out specific words but her tone is reassuringly urgent. I turn back to my father's bedside to see him curled in a fetal position with his foot awkwardly hanging over the edge of the bed. My mother tries to move the leaden limb to a more comfortable position but the weight of it, swollen and lifeless, is too heavy for her. My father tries to turn over, raising his aching body onto one elbow, but he hasn't the strength to see it through. He drops, almost in slow motion, back onto the bed, his bulbous head bouncing on the pillow.

It is not a deliberate action, driven only by gravity with no other force left open to him. The weight of his body forces air from his lungs; a sound that starts as a sigh and grows, as he falls, softening at its crescendo into the distant cry from a wounded beast. No human voice made such a sound; no human voice should ever need to.

The scene before me is frozen by absolute clarity of thought. I am thinking, even as it is happening, that I am watching my father die. That memory is chiseled in granite with a sledgehammer; an inscription visible in places where it has no right to be. It appears frequently, without invitation. It repeats often and I can do nothing but pay attention, fearing its return. Repetition. Attention. Emotion. I know, as surely as I know anything in neuroscience, that this memory will never go away.

2

Man Versus Microbe

Adam
Had 'em.
Anonymous, short poem
'On the antiquity of microbes'*

WE MAKE OUR WAY DOWNSTAIRS TO THE INTENSIVE CARE WARD, PAST THE empty space where Billy's bed stood just moments ago. I catch sight of his quiescent body being wheeled into the elevator, an array of bottles suspended incongruously from his bedstead like optics above a bar, his head directly underneath, mouth open in unknowing optimism.

I smile—partly because it looks so funny but mostly from relief that the pain has been taken away. When the doctors arrived at last they acted quickly; one put a blood-pressure cuff around my father's arm, another stretched an oxygen mask over his swollen head, while a third discreetly pulled the sympathetic, Sixties-style screen around his bed, separating us from all that mattered. It took seconds to sedate him. I wanted them to do it earlier. I wanted the pain to stop, but one nurse said that if my father was sedated they might miss something. Miss *what* exactly? Miss him slipping out for a bit of

---

*I found this poem in the *Oxford Dictionary of Quotations*. Regarded by Google as the world's shortest, it is variously attributed to Strickland Gillilan, Shel Silverstein or Ogden Nash. In the latter case it is entitled 'Fleas', rather than 'On the antiquity of microbes'.

putting practice? Miss him putting the intravenous supply of antibiotics back into his bruised antecubital vein? Miss an opportunity to intervene?

I mustn't let pain-induced aggression get the better of me. Billy is better off now—wherever he is. Great distances have often separated us: when I worked in San Diego it was eleven hours by plane; when he was away on business it could be farther; but I knew where he was and I could talk to him. Now I don't and I can't, and he has never been farther from me.

The Intensive Care Unit's waiting room makes the relatives' room of ward C3 look amateur; the room has a sink and a fridge in addition to the usual low table and half a dozen utilitarian-looking chairs. Exhausted, my mother sinks into their ill-fitting, easy-clean, tan vinyl. The rubber supports have all but disintegrated, retaining less tone than a darts player's abdominals, and her chair can barely support her weight. Acting on automatic pilot, I fill the kettle and look in the cupboard above the sink for mugs. I ignore the cups and octagonal saucers with their cheery pattern of green ribbons and pink bows—too small, too happy for the job they have to do.

A nurse pokes her head around the door. 'Mrs. *Mac*-er-man?' (Actually, it's Mc*Ker*nan, with the stress on the second syllable.)

Instinctively we both nod as she goes on to explain that it will take a bit of time to hook 'William' up to the ventilator and sort out his various monitors (and no one ever calls him William, either). We should get something to eat, and a doctor will be in to see us soon.

I take out two brown mugs and make some tea, sharing one Tetley bag between us. I put a few coins in the accompanying jar, the vestiges of an honor system long ignored by people with bigger problems on their mind than who pays for the milk.

I hate waiting. The worst thing about Intensive Care is the endless waiting. You can smell the cumulative years of people's lives that have evaporated into the small space. The perfume is grief and uncertainty; it's musty, like the expired stock of London train carriages

I remember from my childhood with their dismal maroon uphol-
stery and sackcloth filling. Entering the room is like starting out on a
journey to somewhere—destination unknown. It is inaccessible, behind an intercom-entry system that only adds to the sense of isola-
tion. Life here is transient, separate, suspended. It is like living only
in the present, like living in the world inhabited by the man with no
memory. It is HM City.

On the wall is the most essential item: the public phone, our
only conduit to the real world outside. Seeing it, my mother says:
'We'd better phone the boys.'

The receiver and dialing buttons are grey and sticky from the
sweat of others. I add my own contribution to the cumulative worry,
taking several attempts to dial correctly despite knowing my brothers' phone numbers by heart.

'Billy has been transferred to Intensive Care,' I say, as calmly as I
can manage, conscious that the grubby mouthpiece must have told
the same story many times before.

Ian and Andrew come to the hospital straight away but they
seem out of place. (My older brother, Stuart, who lives in the United
States, is spared the bad news until later, when my mother can talk,
at length, from the comfort of home.) The waiting room is for
sad little people. It's not meant for the likes of my brothers. There's
not enough space between the chairs for two big, healthy rugby players to pace up and down. They can wring their hands like the rest of
us, make cups of tea that aren't drunk, go down to the Crusty Corner for a bacon and Brie baguette and leave it uneaten on the table as
naturally as the other carers, spouses and the about-to-be-bereaved.
But, for the boys, this situation is temporary; the room has no real
power over their lives.

We are reviewing what little we know about our father's condition for the nth time when a man in his early thirties dressed in
baggy hospital 'greens', a hair net and chunky white clogs slips in
from an adjoining room and introduces himself as Dr. Uli, the senior registrar.

'You are all William McKernan's family?' he asks, reassuringly pronouncing our surname correctly.

'Bill McKernan,' my mother says. 'I call him Bill.'

Dr. Uli shakes our hands as we introduce ourselves in turn.

'Well, your husband,' he says to my mother, 'and your father,' he says, looking straight at me with a solemnity equaled only by the priest at my wedding, 'has a serious infection, probably bacterial, which has got into his bloodstream and made him very ill. His body has gone into septic shock.'

Septic shock? Isn't that where blood pressure drops so dramatically that the body's organs are starved of oxygen and seize up? How can bacteria do that? How can something so bloody small be so brutal? Bacteria are incredibly, insignificantly small. You could stamp on billions of them with your big toe alone. They're so tiny we didn't know they existed until 1674. We didn't even know that there could be bacteria in your body until the day Anton van Leeuwenhoek examined his own saliva and urine with his homemade microscope. It's worryingly amazing what's to be found below the limit of human vision once you're able to look close enough.

'We think the infection started just underneath your husband's eye,' Dr. Uli continues.

Could it have been caused by his habit of scratching the bags under his eyes? How did something so insignificant get my father in this state?

I know that, in general, microbes aren't harmful. Fewer than 1 percent of them have the power to infect man and we go about our normal business untroubled by what is breeding in the dirt around us. We give no thought to the plaque-forming bacteria between our teeth or the diverse fauna in our intestines. Microbes give us wine, bread and cheese, and we let them decompose our bodies when we die. It seems a fair deal. When Louis Pasteur, the French chemist and microbiologist, first proposed that bacteria live on animal matter, competing with one another until they take it over completely in decomposition, few believed him. Even in Florence Nightingale's time, nurses did not believe that bacteria cause disease. Less than two hundred years ago the accepted dogma was that disease produced bacteria, not the other way round. On the whole, we well-developed mammals have ways of keeping small unicellular organisms in their place. But sometimes they get further into our bodies than we would want and, when that happens, the battle of man versus microbe begins.

'We have taken swabs to try to identify which kind of bacteria is causing the problem. Once we know that, we can be sure we have picked the best antibiotic. We have also given your father fluids to keep his blood pressure up.'

At my request, Dr. Uli itemizes the list of drugs that have been injected into his bloodstream. He counts them off on the fingers of two hands. The pharmaceutical interventions speak for themselves and take me back to my undergraduate pharmacology class. My group's first practical attempt to control the blood pressure of an anaesthetized guinea pig was pitiful in its ineptitude. Despite our cautious and well-reasoned efforts, the chart we produced looked like a cross-section of Kent, rising and falling over the North and South Downs before dropping precipitously at the white cliffs of Dover into cardiac arrest.

Dr. Uli asks if we would consider including our father in a clinical trial of a new antisepsis treatment. The hospital is taking part in a multicenter experiment to test a treatment using antithrombin III, one of the body's natural anticoagulants. Along with a couple of other equally complex-sounding factors, its normal job is to prevent blood from clotting in the tissues. (Why is his blood doing that? Why isn't it in his blood vessels where it belongs?) He explains that supplementing the body's natural anticoagulants may improve Billy's chances; it should help maintain his circulation and perhaps reduce the mortality rate. My mother nods as if she has understood the explanation but I can see from the look on her face that she hasn't translated the phrase 'reduce the mortality rate' into 'is less likely to die'.

'What do you think, Ruth?'

'Is it extracted from human blood?' I ask.

'I believe so,' he says.

Do we want to do this? I know too much and too little. Anything extracted from human blood could contain prions, the particles responsible for new variant Creutzfeldt-Jakob disease (CJD). How do they extract antithrombin III? Would prions co-purify? I'd need to know. Is donated human blood screened for prions? No, can't be: no test has been invented yet—and, even if it had, the level of prions could be very low and still be infective. After all, prions are incredibly hard to destroy. If any of the blood was contaminated we could be agreeing to inject dangerous particles straight into my father's bloodstream. How long would it take them to get into his brain? How long would it take for him to die of the human equivalent of mad cow disease? Are we helping him to recover only to infect him with something worse? No, hang on: the incubation time for Variant Creutzfeldt-Jakob disease (vCJD) is years. Even if Billy recovers from this he's already in his sixties. What are the chances that he's going to contract CJD and die of it before something else gets him? Get a grip. You're in a mental cul-de-sac—it really doesn't matter.

'Your father was always prepared to take part in medical experiments when he was a student. He had no end of radioactive molecules injected into his body when we lived in Birmingham. I'm sure he would approve,' my mother says. So we agree. We sign the forms and add my father's name to the terabytes of data that such a trial generates. (I never find out whether my father is infused with antithrombin III or placebo, but much later, in 2002, I find the results of the study. It is a huge trial involving 2,314 patients from Intensive Care Units all over the world. Working in a pharmaceutical company, I know that the cost of such a study must run into many millions of pounds. The outcome of the trial is clear: close to 40 percent of all those enrolled died, irrespective of whether they were given antithrombin III or not. The experiment was successful in that the study gave a clear result: antithrombin III didn't benefit the patient at all; but the experiment was properly done and at least clinicians now know it won't help and can start looking for something else. So far, that something else has been frustratingly elusive. In the last decade, billions have been spent on more than seventy clinical trials to find a way of treating sepsis with only one new therapy making it to market.)

Ian asks whether finding out which bacteria it is will help. I know it will. We don't need to waste time with questions of this type. Somehow, I hadn't expected my brothers to ask questions; it hadn't occurred to me they might be as desperate for information as I am. Ian is an industrial chemist without much biological background. My father isn't just their father but their boss, CEO of their company, ultimate source of business information and advice. Ian and Andrew have even more to lose than I do.

'Maybe,' says Dr. Uli. 'But we are giving him a broad-spectrum antibiotic which is effective against most of the likely bacteria.'

How can something one-tenth the size of an average blood cell wreak so much havoc, dodge my father's natural defenses, destroy great chunks of flesh and leave a trail of destruction through the body on a scale matched only by the last fifteen minutes of a high-budget Bond film?

It's all to do with evolution and refinement. Microbes have been around since the primordial soup was simmering. The course of their evolution has recently become clearer. In the late 1960s, Carl Woese from the University of Illinois used molecular genetics to compare similar segments of DNA from hundreds of organisms, allowing him to work out how they are related. Now microbiologists can track entire family histories of microbes, understand when new features emerged, and you just can't help but admire the panoply of skills bacteria have picked up along the evolutionary way.

Genetic changes occur when cells divide, and evolution is linked to the rate at which an animal reproduces. In which case, bacteria can react to a changing environment at incredible speed because it takes them, on average, just twenty minutes to divide. They are evolving one and a half million times faster than we are. 'So why are they still around as bacteria?' you might ask. 'If they are so good at evolving, how come they haven't evolved into something else?' And of course the answer is that they have but they have also managed to keep their original unicellular place in nature too. Some don't seem to have evolved at all. The earliest strain of cyanobacteria that lived on Earth 3.5 billion years ago can still be seen today off the northwest coast of Australia as fluffy, cauliflower-like colonies called stromatolites.

Single-celled organisms have filled niches that would leave Galapagos finches with their beaks wide open in awe. They have evolved to live in high and low temperatures, to harvest energy from the Sun by photosynthesis and to live on many different chemicals including sulphur, oxygen, methane and crude oil. Other bacteria have evolved into smaller parts of larger beings. Millions of years ago purple bacteria became mitochondria, the sub-cellular factories present in almost every cell in our bodies that produce the energy we need to survive.

Not only have whole categories of bacteria evolved into different species, they have also been more than creative in other directions too. When DNA duplicates, errors occur at a reasonably constant rate, and we now know that each time microbes divide between one and ten mutations occur in their DNA. All errors are tiny and most are inconsequential. Some, however, are detrimental and slow down the microbes' growth so that mutant populations get overgrown by tougher strains. However, in a selective environment, such as in the presence of an antibiotic, any rare cell that manages to carry on dividing will flourish, while the original dies out. This awesome capacity of bacteria to cope with an ever-changing environment, and their ability occasionally to exchange bits of DNA as they divide, allowing them to pass on newly acquired properties to their neighbors, keeps them one step ahead of the rest of us. Based on the appearance of ever more virulent new strains, some might argue that they are moving farther ahead all the time. The rapid emergence of antibiotic-resistant bacteria in hospitals throughout the developed world illustrates just how flexible and adaptable microbes can be.

But we have weapons too: twentieth-century medicine and the combined expertise of Addenbrooke's Intensive Care team. This is one of the best teaching hospitals in the country; Cambridge is a center of excellence for all things medical. I try not to think of my father as an assemblage of different biological processes, as the doctors here surely must. To me, he is more than the categories of his physiological problems, more than an object in the hands of mere mortals. Dr. Uli displays the accoutrements of rank and authority: an electronic pass in the crevice of his V-necked tunic, a weighty bunch of at least a dozen shiny keys dangling from his hip and the obligatory stethoscope, worn in this year's style, casually slung around his neck like a scarf. But can I trust him, just because he is a doctor? I've worked with medics before, in the lab, and I've witnessed some of their mistakes. I've seen the infusion pumps set incorrectly, the drug dilutions that were mistakenly calculated tenfold out. And only now do I see the bits of human tissue that I once witnessed being homogenized and accidentally thrown down the sink as part of someone else's father.

I try not to think what the medical profession might be doing somewhere on the other side of the wall. I can dissect Dr. Uli's words as much as I want, check the facts he gives us in medical textbooks, as I choose to do, but it will not change my father's condition, the complexity of which our adviser reduces to two painfully simple phrases: 'This is a very serious infection. Septicemia can be fatal.'

Ian is motionless. Andrew lifts his spectacles and drags his hand down his forehead, wiping his eyes with his thumb and forefinger. He stares at the ground. Their stillness is alien to me. Andrew hasn't sat still in one place for more than thirty minutes since he left school. When did Ian become so adult, so controlled and worldly-wise? Having seen my brothers only at family holidays, dinners and social events, I believed them to be cavalier live-wires constantly joshing, arguing, larking around. This morning I have learned that they are grown-ups, too.

'Poor, poor Billy' is all my mother can say.

So much for my initial judgement that infection would be the most favorable option! Statistics show that a third of people with sepsis don't make it to the end of the month, an outcome that rivals myocardial infarction (heart attack) and is worse than most common cancers. If discussion of the clinical trial didn't make it clear enough, we certainly understand now.

While we stand there helpless, the battle continues between the growing population of bacteria and whatever defenses my father's body, with the help of twentieth-century medicine, can muster. If the bacteria merely penetrated the body, as in an ordinary sore throat, ear infection or conjunctivitis, there would be no catastrophe. They might sneak under the skin and breed into a small colony, but white blood cells would soon 'seek out and destroy'. In the process, antibodies would be generated to prime the immune system in readiness for any subsequent invasion. No lasting harm would be done and the victim would be back on the golf course quicker than you could say 'Staphylococcus aureus'.

However, some bacteria produce toxins that interfere with normal cellular function, while others promote their own colonization by neutralizing the host's defenses. The most advanced bacteria have evolved many ways of fending off the mammalian immune system. Some have a thick outer capsule made of hyaluronic acid—microbial Teflon—which resists attack and degradation. Some have strings of sticky Velcro-like molecules that help the bacteria attach to their surroundings, making it physically difficult for white blood cells to surround them. Phagocytosis—the way white blood cells destroy bacteria by engulfing and digesting them—requires 360-degree access. Bacterial defensive strategies read like a historical evolution of combat techniques. They range from medieval tar on the path—the secretion of extracellular adhesion proteins that prevent white blood cells approaching—to Second World War radar-blocking signals—secreted enzymes that dissolve white blood cells' distress signals—preventing additional reserves from being recruited. Bacterial tactics continue to improve. If a new characteristic offers bacteria a competitive advantage it inevitably gets added to the armory, whatever the consequence for its human host.

'Dr. Park, the consultant of our ICU, would be happy to see you later this afternoon when we have more information,' says Dr. Uli, as a parting comment. 'In the meantime, it is OK for you to go into the unit to see Mr. McKernan if you'd like to. Preferably only one at a time, though—you'll see that there isn't much room in there.'

There are times when you just *know*, when something is so obvious it needs no discussion. One of us has to go in first and I know it's me. Normally, I would feel no fear going into an Intensive Care Unit. Whatever lies through the porthole window holds no mystique. I am fortunate, if 'fortunate' is the appropriate term, that I have visited such a ward several times before. I occasionally helped collect organs for transplant or research during my two-year stint at St. Mary's Hospital in London. I've seen the insides of a human body close up and personal. I know how precisely the tissues fit together and how much stuff you have to displace to find the kidneys. I know that there is quite a bit of variety in there; we don't all look the same on the inside either. Rather more relevant to the present situation, I once visited an asthmatic friend from university who died of anaphylactic shock. If any benefit was ever to come from that traumatic experience, perhaps it was the preparation for this one.

PLEASE WASH YOUR HANDS! I know not to dodge past the sink that guards the entrance. Conscientiously, I follow the instructions on the wall. I squirt the pink liquid into my palms, turn on the water by flicking the silver levers with my elbows and then pass my hands alternately through the scorchingly hot and ineffectively tepid-cold water. There must be better ways to limit the spread of bacteria in a hospital. In the lab we would squirt alcohol onto our hands, rub them together and let them dry inside the air hood when we did experiments under sterile conditions. Now we wear disposable gloves, but of course they add to the cost. It might be difficult to prevent superbugs from arising—the pull of evolution is a powerful force—but we could surely limit their spread from visitor to loved one, from patient to patient. A few simple experiments could tell us how to minimize bacterial contamination here. My mind wanders into the realm of experimental design. But thoughts of microscopes, samples of bacteriocidal agents, agar plates and fingerprints are soon displaced by more important things.

The background sounds in Intensive Care are not like those in any normal hospital ward. The atmosphere is unfamiliar, the tempo slow; no bustle, no chatter, no banter. There is no one for the nurses to supply with bedpans or acetaminophen or cups of tea on demand. There is no need for chipper auxiliaries to jolly life along. Care is serious, industrious and mostly meted out to the unconscious. A nurse attends each occupied bed and several more sit at a long desk. Behind them a big white board lists the patients' names. I scan past those I don't recognize, down to Bed no. 4, next to which is written: '1347065 William McKernan'. Alongside the name is his date of birth and age: there can be no confusion.

I sit down by my father's bedside and, despite the constant sub-dued activity, it is a frighteningly peaceful, almost musical, place. More electronic gadgetry surrounds Billy than I've seen in the best-stocked hi-fi shops of London's Tottenham Court Road. He is fully wired for bodily functions, and what life remains inside him is translated into sounds and lights on the surrounding hardware. His unique rhythm, made public, establishes the cadence in his corner of the ward. A respirator sets the droning background rhythm; metal boxes monitor a different beat; in between the bleeps I see the numbers flash, I watch the digits change and hear a patient sigh.

My father looks relaxed, lying still, with a yellow cotton blanket folded neatly below his armpits and tucked under his feet. Intravenous propofol has shut down his brain and put a final end to the frenzy. Yesterday my mother held the hand of a ranting, distressed psychotic while the nurse slipped in the first of many plastic lines. It feels as though that was another man in another life on the other side of the world. Now look at him. The respirator tubing, loosely held in his mouth by tape, hangs over his lower lip like one of the cigarettes he gave up thirty years ago. Thoughtfully, his hair has been combed and, even though his face is still swollen and red, he looks as if he's sleeping, not dying. Seeing him sleeping is familiar, comforting, although usually it's after dinner, in front of the television, with an empty wineglass on the side table.

The bag under his left eye has become black and crusty, like burnt toast. The black triangle marks the temporary entrance through which the bacteria entered and my father left. I pick up his hand and hold it: it seems the right thing to do. I'm not prepared for the coldness of his fingers. Those tough old fingernails are tinged blue, a hue that virtually extends up to his wrist. Automatically I curl my warm hands around his and gently rub the back of his hand, in the same way that I have rubbed Adam's after he's been playing in the snow.

'A side effect of the drugs,' says the nurse stationed at the foot of his bed, and he carries on with his duties, explaining what he's doing as he does it.

I'm not quite sure whether his comments are reflexive or meant for me, so I nod anyway, just in case. As he notes down a reading I surreptitiously scrutinize the charts hanging over the end of the bed. Joined-up dots show that my father's blood pressure was normal on admission and that it has steadily dropped over the past twenty or so hours. I survey the equipment and manage to pick out the blood-pressure monitor from the rest. It records ninety over fifty.

'Is that enough?' I ask.

'Better than it has been,' he replies.

Momentarily all is under control. Then systolic drops to 89, then down to 88. My chest tightens. Back to 89, then up to 90 again. I breathe a little easier. His blood pressure is driving my physi-ology. Such fluctuations feel unpleasant, yet I'm content. I'd like to stay, be part of the picture, but others are waiting for news outside.

My mother is talking to a surprisingly cheerful, solid-looking woman when I return. Isobel is Scottish, sixtyish, wears good walk-ing shoes, a quality wool suit and a Liberty's peacock-patterned head-square. She is visiting her husband, who has been in the hospital since his aorta ruptured nine days ago. Isobel watched him disappear into the operating theater. She kissed him, told him she loved him and waited until he passed into Intensive Care, where he has yet to emerge. She educates my mother in the nitty-gritty of established routine. 'Don't come too early,' she says. 'You can't get in while they're doing ward rounds. Afternoons are good and the best chance of see-ing a doctor is early evening.' As she chats, Isobel takes out her sew-ing, threads some yellow ribbon into a huge bodkin and carefully adds a few more petals to the growing bowl of roses on her fabric. She does it tenderly, reminding me more of Ugolin sewing Manon's ribbon to his heart in the film *Manon des Sources* than of more jo-vially tying it round the old oak tree.

My mother accepts that it's her turn to go in. Anything is better than shifting about in ignorance on the tan vinyl cushions.

'It's OK, Mum. You'll be fine, really. He looks much better,' I say, fully aware of how ridiculous this sounds.

For Ian and Andrew, the choice is harder. Ian saw Billy the previous evening, so he knows what to expect, but Andrew, whose last memory of our father was at work and unsuspectingly well, is reluctant. Why supplant that image with something worse?

'It's up to you, son,' says my mother supportively. 'He won't know, he won't mind.'

Maybe later—if things look better. Maybe he'll go in then.

I'd rather sit there than sit here, any time. It is comforting to just watch him, as I used to watch my children sleep when they were small. No harm could come to them while I was there.

From behind the desk, piled high with medical textbooks, papers and files of patients' notes, Dr. Park doesn't paint too optimistic a picture. He carefully balances positive and negative comments in equal measure, adding in turn to either side of the scales. They have identified the bacterial cause of infection as a streptococcus. (How did they do that? Culture a sample of exudate from under his eye? An antibody test? Cell surface markers visualized under a microscope? Polymerase chain reaction of its DNA?) Billy has an underlying blood disorder that means his immune system is inefficient and so his ability to fight infection is reduced. However, despite his recent medical history, he has been a fit, strong and healthy man for most of his life and he took regular exercise, so that goes in his favor.

Dr. Park's prognosis is neatly packaged in bite-sized chunks and spoken with the practice and confidence of a man who has served up the same phrases many times. How can doctors sound so sure? Scientists do the research that underpins medical confidence. Yet we rarely believe something fully. There is always space for a different interpretation or a reappraisal in light of new facts. But somehow, in translation, absolute facts emerge and Dr. Park delivers them sensitively and with solid belief. He is in a critical but stable condition. It is too early to say. The next twenty-four to forty-eight hours will be crucial. If Billy can get through this he has a chance.

'Go home and get some rest,' he says. There is nothing we can do but prepare ourselves. If Billy recovers we will need our strength to help him, and if he doesn't we will need our strength to help one another.

Streptococcus! At least the enemy has a name and a pretty unpleasant-sounding one at that. *Streptococcus* is not just one type of bacterium but a whole family of bad players, the first of which, *Streptococcus pyogenes*, was described in 1879 by Louis Pasteur as a 'chain of beads'. The same class of bacteria, albeit secreting different toxins, was responsible for the epidemic of scarlet fever in Europe in the 1800s, the outbreak of rheumatic fever in the United States after the Second World War and the recent dramatic rise in toxic shock from Saffron Walden to Santa Fe.

Bacteria are not just defensive. Not only have they acquired properties to evade destruction, they can release an offensive range of factors too. Different strains secrete a different collection of biological molecules or toxins. While some rupture red blood cells, others produce fever. In fact streptococcus is the Joan Rivers of bacteria. Just when you think that every possible vitriolic tack has been taken, lo and behold, out it comes with a new one. And even though you might not appreciate exactly what it does in the world, you still can't stop yourself from admiring its versatility and skill.

Streptococcus's most recent tactic is to disrupt the circulatory system directly. The 'super-antigens' it produces hone in on a specific type of white blood cell, the T-lymphocyte, with more precision than a scud missile. Normally blood vessels dilate during infection, allowing white blood cells access to fight the invasion, an exercise not unlike clearing the streets to let the army through. Activating just one T-lymphocyte in 10,000 is all that's required. But in streptococcal toxic shock up to 20 percent of T-lymphocytes are ruptured, spewing out excessive vasodilatory and inflammatory molecules. Blood pressure drops dramatically, fluid leaches into surrounding tissues, clots form in the tiniest blood vessels and ultimately the circulatory system collapses; a process known as toxic shock. Scientists understand that the exploding T-cells do the real damage because a strain of mice, lacking T-lymphocytes, is resistant to bacterially induced toxic shock, but if these animals are transfused with T-cells from normal mice: hey presto! they go into toxic shock, just like their cousins.

We sit around the little table, flicking through month-old Sunday supplements and veteran editions of *Country Life*. When we've had enough we just look at one another. We've seen Isobel come and go. We've seen members of three other families pass through, but mostly they don't stay. They deposit their coats and low-value belongings while they visit their loved ones. Then, just as efficiently, they collect their things and leave. We are the new blood, the objects of empathy from the more experienced. Uninitiated, we have brought no books to read, no embroidery, no pocketfuls of coins or fresh milk. We sit there because we don't know what else to do. We discuss what we should do, as though in a position to make decisions. Should we stay a little longer, just in case? I'll check one last time that there is no change and then, as Dr. Park suggested, we'll all go home, even though it seems like we're deserting our father in his hour of greatest need.

I am barely through the door into Intensive Care when my whole body stalls. My limbs refuse to move. The information from my brain can't get out to the rest of my body. Only one cerebral department is not on all-out strike and it has a single train of thought: let him die. Everyone else wills our father to live. What a traitor I am, frozen with guilt. Let him die. Please, let him die. If he's going to die let it be now, not after days of hopeful vigil. I don't want my mother to be Isobel. I don't want to be Isobel.

I feel my pulse throbbing in my temples, salt stinging my cheeks. I grab some coarse blue paper from the roll above the sink, drag it across my eyes and wait a few moments for the vasodilation to subside. I check my face in the mirrored tile above the sink to be sure that the redness is no longer obvious and return to my mother and brothers.

'He looks just as he did before,' I say. 'There is no change.'

So we leave our phone numbers with the sister on duty along with the instructions that if anything happens she is to call my mother first and then the rest of us.

'Call me too,' I say. 'Call me too, please.' And we leave to spend the night in constant fear that the phone *will* ring.

Later, I find a poem, 'Back' by Al Alvarez, some lines of which perfectly capture the sense of shock.

> *The night I came back from the hospital, scarcely*
> *Knowing what had happened or when,*
> *I went through the whole performance again in my dreams.*
> *Three times—in a dance, in a chase and in something*
> *Now lost—my body was seized and shaken*
> *Till my jaws swung loose, my eyes were almost out*
> *And my trunk was stunned and stretched with a vibration*
> *Sharper than fear, closer than pain.*

With the benefit of hindsight, and my father's unusual symptoms notwithstanding, everything begins to fit, such that, in the words of Sherlock Holmes, 'a hypothesis gradually becomes a solution'. Once I have done a substantial amount of reading, it becomes clear to me that everything my father endured can be accounted for by a bacterial infection. All the symptoms, the trauma, the whole gamut of bodily catastrophes, could be laid at the door of the dishonorable microbe and explained by 'streptococcal toxic shock', a cascade of events that happens so frequently these days that it has officially become 'a syndrome'. There were 2,000 or so victims of Streptococcal Toxic Shock syndrome in the United Kingdom in 1998 and the number of cases has steadily climbed.

On admission to Addenbrooke's Billy became not just a patient, but a contribution to medical statistics. He is a case, a human data point; one of the 20 percent of patients in whom an early symptom of streptococcal infection is fever. That would have been the suspected flu for which he took two acetaminophens and went to bed. He developed nausea, vomiting and diarrhea and, judging by his inability to respond to my questions as he wandered, distraught, through the ward when he was first admitted, he would also be included in the 55 percent who become confused or disorientated because the brain receives insufficient glucose or oxygen. Streptococcus infects soft tissues, such as the skin, causing cellulitis. The exceptional pain of cellulitis is one of its distinguishing features. It can be so bad that even the weight of a sheet or light pair of pajamas can be unbearable, explaining my father's uncharacteristic need to shed his hospital gown in public.

It is rare for septicemia to develop from a seemingly innocent scratch under the eye and progress with such speed. I know because I scour the scientific literature. Normally clinical papers make dry reading. They talk about cohorts of patients in large numbers. The size of each group depends on how many people are needed to show a statistically significant effect. If a large or easily measurable difference is expected, then fewer patients are tested. For drugs that lower glucose in the blood of diabetics, only a handful of subjects are needed to demonstrate a robust benefit, whereas an antidepressant trial might have seventy patients in each group. I find only one paper that describes toxic shock starting as an infection around the eye, and it refers to just one single man, not a collective at all. The author was an ophthalmologist from Chicago who wrote about a sixty-two-year-old who rapidly progressed into toxic shock and multi-organ failure after a hemolytic strain of bacteria infected the small cut in his eyebrow. When a paper is written about just one person, it has to be something extraordinary, such as the legendary story of HM, otherwise it would be dismissed as unscientific. The publication in the *American Journal of Ophthalmology* in 1997 tells me that Billy's experience is a rare occurrence indeed.

Medical statistics give the odds of a man of my father's age with my father's medical history surviving streptococcal toxic shock with multiple organ failure as no greater than one in three. As a scientist, I can appreciate the qualities of the proliferating colony of streptococci, packing an artillery honed by millions of years of evolution. I even have some admiration for its evolving credentials. In retaliation we have only a few hundred years of modern medicine to offer.

Bacteria that gain access to the store cupboard in our blood use the glucose, amino acids and energy derived from them to multiply. When circulating stores of nutrients are exhausted, they can be supplemented from the nearest flesh if bacteria force the body to break down further. Glycogen from the liver is turned into glucose, and muscle is reduced to its elementary amino acids. Physics 101 tells us that matter is neither created nor destroyed in the course of a chemical reaction. Building bacteria are no more than a series of chemical reactions, and an expanding colony is built with raw materials stolen from the host's flesh.

From the microbe's viewpoint, Billy's body is a tub of nutrients, a man-sized culture dish, replete with everything it needs to grow and multiply. And all the bacteria have to do are use the abilities they have evolved. The expanding colony is the estate agent of the 1980s or the e-commerce broker of the 1990s. It can't fail. The conditions are right; the raw materials are all there. It has everything it needs to make a killing.

3

Consciousness

. . . the inner space
Of private ownership, the place
That each of us is forced to own,
Like his own life from which it's grown,
The landscape of his will and need
Where he is sovereign indeed.
From 'New Year Letter' (1940) by W. H. Auden

CONSCIOUSNESS, LIKE HAPPINESS, MEANS DIFFERENT THINGS TO different people. Right now, it's the fundamental property that separates my father from the rest of us. What is it, this elusive quality? Animal or vegetable or mineral? For centuries, man has been trying to understand the curious quality that might separate us from some other species. Consciousness has been described as spirit, religion, a foundation, the biggest white space on the map of human knowledge. It has been compared to a stage, a field, ripples in a pond. The doors are open for debate. Everyone gets to have an opinion on what it is and—just as importantly—what it isn't. The good thing is that we are so far from any real understanding that no one can be wrong. Philosophers, writers and scientists all agree on one thing, though: definition is difficult.

Virginia Woolf described consciousness as 'the structure on which life's luminous halo is built'. The sacred Hindu scriptures or 'Brahman' of 800 BC considered consciousness to be 'the supreme mystery beyond thought', while the philosopher Thomas Nagel believed that we can never know. The subjective experience of consciousness can never be attained through objective science, he thought, and to demonstrate its unknowable quality he posed the unanswerable question 'What's it like to be a bat?' David Lodge, a scholarly 'priest' of the humanities, considers the nature of consciousness from both scientific and literary perspectives in his recent book, *Thinks.* His characters set out to answer Nagel's question in the literary style of Martin Amis, Irvine Welsh, Samuel Beckett and Salman Rushdie! Amis's bat hangs upside down and craps a lot; Welsh's bat scours the countryside for a bloody fix and swears a lot; Beckett's bat feels darkness with a blind man's sensitivity and squeaks a lot; whereas Rushdie's bat questions, in true Rushdie style, what it's like to be a man.

So let me have a go at defining what consciousness is. I'm entitled to comment, to mix my metaphors with everyone else's, because I've been obsessed by it, searching for it, waiting for any sign of it to return in my father. In the same way that negative space defines a Henry Moore sculpture, the inexpert observer recognizes consciousness by what it isn't, rather than by what it is. And I have been studying its absence in careful, anxious detail.

I marked off the hours and willed myself to sleep last night, just to make time pass. By six o'clock I gave up and called the Intensive Care Unit for news. Hesitatingly, I called my mother, knowing that she would dread the sound of the phone. But I need not have worried: while I had waited until six, she did well to wait till five. And now I'm back here at his bedside because, well, what's the point of being anywhere else?

I'm staring at the consistent lack of expression on my father's empty face. He is totally unconscious. There is no jot of emotion, no pain, no frustration—no frown, no muscle tension at all. Being unconscious puts him in a state of unremitting peace, like a Morandi still life—reality with the color wrung out—or a silent 1950s home-video frozen in one restful black-and-white frame.

If you had asked me a few days ago, I would have prattled on knowledgeably about consciousness being the cutting edge of research. Now that religion and science cohabit comfortably in most societies (with a few notable exceptions such as the creationist strongholds of the southern states of the United States, where evolution is seen as a blasphemous theory and is forbidden to be taught in schools), studying consciousness is one of the most exciting things going on in neuroscience today. It is now academically acceptable to work on the enigma of consciousness without being branded a heretic or heathen. For most of this century, scientists were reluctant; consciousness was a scientific taboo, the domain of philosophers. Before my father's illness I would have said that the topic is ripe for research, back in vogue, a legitimate problem of biology and physics, like any other. But now, reducing it to a question of science is too impersonal to bear. It's like watching a man drown while only reflecting on how hydrogen bonds make the water molecules around him liquid rather than gas.

It is noisily quiet in here with no human sound between the bleeps and clicks of cardiac monitors, infusion pumps and respirators. There is nothing else to do but think: think about how little we understand of what distances my father's state so far from my own. Scientists are still at the stage of developing analogies and homing in on descriptions that might direct us to testable theories. We understand as much about consciousness as the nineteenth-century monk Gregor Mendel knew about modern-day genetics from his endless studies on peas. The inherited characteristics of smooth and wrinkled varieties of peas made scientists think about what governs inheritance, but understanding its molecular nature required a whole revolution in biochemistry and physics. We know no more about consciousness than Galileo knew of black holes when he made his first telescope in 1609. The chances of a twenty-first-century scientist understanding consciousness are as likely as a pre-Raphaelite blacksmith deciphering how a mobile phone works. We don't know what tools we'll need or even if we have them yet. Will we have to wait for the next revolution in electrophysiology or particle physics to understand the nature of consciousness?

In the absence of real knowledge we do what Mendel did: we observe, describe in detail what we see. There is something public about consciousness—it is our gateway to the external world. We can make sense of what's around us and respond appropriately, a property we arguably share with most highly evolved members of the animal kingdom. This isn't so for the man in front of me, obviously. But consciousness doesn't just mean awake and responsive; it describes a higher level of sensation. We are not just conscious, we are conscious *of something*, of whatever we are thinking and feeling at this very moment. Such sensations allow there to be something very private about consciousness, too. My thoughts are not available to others, unless I choose to make them so. Are feelings consistent? we ask. Is the pain I feel exactly the same pain felt by my mother or anyone else in the same situation? Such qualitative experiences, or qualia, are a philosopher's puzzle. What is the objective feeling of pain, the taste of champagne, the experience of seeing red? Does everyone see and feel things the same way? Is it possible even to convey our exact feelings and their meanings to others? The tantalizing hope of doing just that, of re-creating a cerebral sensation using words or images, has kept artists and writers in business for centuries.

His hands are so cold. Maybe I should bring in the blue Icelandic thick wool gloves that my mother knitted. She made matching hats, gloves and scarves for the entire party that went skiing to celebrate his fiftieth birthday. My father wore his when he and my mother won the downhill sledding competition—Les Diablerets slalom, we called it—a victory he embellished with boyish boasts about their superior tactics, although their success could more fairly be attributed to my parents' greater combined mass and the incontrovertible laws of gravity.

That thin polypropylene tube disappearing into the back of his hand carries the molecules that keep him alive. See the needle held in place by a strip of translucent tape, like a strip of airmail paper— like the letter my father wrote from Singapore wishing me a happy twenty-first birthday. . . . Where is that letter? I'd like to read his welcome to adulthood again—and, as I remember, there was a good line about cockroaches in his bed. As my mother had instructed him not to share a room with any strange females and he didn't know how to tell the sex of a cockroach, he thought he should move to another room, to be on the safe side. It was just a couple of pages of forward-slanting, spiky scrawl that was completely him and completely personal without saying anything profound.

We call what goes through one's head at any one time a stream of consciousness. It has a very limited temporal capacity. What we're thinking is only a tiny fraction of the brain's content, just as what's available on the screen of a laptop is only a tiny fraction of the hard disk. A stream of text rolls past our attention, roving from file to file, using any type of link available.

The analogy of the brain as a computer is one way to generate questions and hypotheses. Screens, monitors, wiring, tubes: it's all hardware. If neurons form the brain's hardware, can studying them ever reveal the information they contain? If you take a computer apart where do you find the Simms? Scientists accept that consciousness is a product of the brain and, unlike the dualists Descartes or Samuel Johnson, who made a clear distinction between brain and mind, we accept that the brain is *all* you need. As Somerset Maugham said, 'The highest activities of consciousness have their origins in physical occurrences of the brain just as the loveliest melodies are not too sublime to be expressed by notes.'

In this tuneless little corner of the world around my father's bed, artificial instruments provide the meter, an empty beat on top of which no melody is played. The basics are there, made by his body's rhythm—heart and lungs—outside, on view. I don't think my father would like having his physiology displayed like this, on screen, exposed for all to see. The yellow light on the cardiac monitor follows an atypical but regular four-beat bar. Peak, trough, wave fades. Light dims left to right. His artificial lungs add a complex lower sound. I'm hoping for, half expecting, something else to add in. But this isn't an introduction to the orchestra or intro and outro by the Bonzo Dog Dooh Dah Band. There is no other melody, just an empty stave where bars and lines are stretched out into an ill-defined future by a team of mostly anonymous men in thin green cotton tunics.

We can describe the brain as a cardiologist might view the heart: a functional organ tied to the environment, tied to circumstance. Patterns of activity describe the operational state of both organs. The way in which the heart beats faster when we run can be described by the number of beats it makes in a second, the volume of blood that passes with each beat, the level of adrenaline and oxygen in that blood, the force of contraction and so on. The way that the brain works can be described in terms of physics and mathematics too. Its function can be measured by the amount of oxygen and glucose it uses, and the frequency and form of its electrical activity.

When we are awake, brain cells send electrical signals to one another with a regular rhythm. While the heart beats about once a second, neurons fire with bursts of activity about forty times faster than that. Only a small proportion of cells fire each time, but the rhythmicity is there, like the pulsatile flashing of Christmas tree lights. At first they might seem random, but there must be a controlled sequence within them. Think of the flashing lights in New York's Times Square. The last time I was there, a panel of small red lights displayed current news and weather. Patterns of lights were illuminated to spell letters. Letters moved across the board in sequence to form words, then short phrases, which I could read as the news headlines of the day. 'Mets win again. Progress in Middle East conflict. Outlook sunny.' Orchestration of the whole pattern of lights switching on and off was necessary to relay the news meaningfully. Synchronization was vital to convey meaning. The brain seems to operate similarly with patterns of activity that have a characteristic frequency; consciousness is somehow linked to neuronal activity at forty cycles a second.

Defining consciousness as brain activity in the range of 40Hz is just a start. It is the simplest physical description of the processes that underlie a complex range of meanings and sentiments. It is rather like describing a radio channel as the frequency at which it operates. It tells us nothing about its contents or style, whether it is transmitting music or news. The radio frequency carries the message, but it is not the message itself. As with the radio, we cannot tell a person's conscious thoughts from the firing pattern of the brain. There are clues of association. If cortical activity increases, that would suggest thinking, analysis and reasoning, but we cannot read the details from any known measurement of brain activity.

Shona, Billy's current nurse, moves over from the adjacent patient's bedside. She takes a small glass vial from the trolley at her side, breaks it open and sucks the milky fluid into a 2ml disposable syringe. The diminutive Scot opens the green three-way tap resting on my father's forearm and gradually injects the contents. Another shot of propofol. Shona looks at each monitor in turn and adds a few more words to the notes. Hers is the third type of handwriting on the page but I am too far away to read any of it.

What would happen if he wasn't given propofol? Would he still be unconscious or do the doctors keep him like that so they can treat him more effectively? Consciousness is controllable by man—or, at least, by drugs—to the extent that we can choose to remove it. When lightly sedated with a high dose of Valium, the firing frequency of brain cells is reduced from forty bursts a second to about twenty-five. Fully anaesthetic drugs, like propofol, the intravenous anaesthetic of choice in Intensive Care medicine, slow activity even further, to maybe twelve or fifteen bursts a second.

Below a threshold level of twenty to twenty-five bursts a second, the brain can no longer support consciousness.

Something in the synchronous firing of neurons allows consciousness to emerge. Even the poet Geoffrey Lehmann could imagine that 'the night sky of the mind allows a single thought to light up as a sentence'.

For my father the night is starless. I check the screens reporting his various physical measures. Patterns of electrical activity in the brain can be monitored by electroencephalography (EEG). Electrodes placed on the skull record the combined activity of large cerebral areas, simplistically described as brainwaves, which have a characteristic form depending on whether the subject is awake, asleep, dreaming or unconscious. Why is there no EEG monitor? Perhaps what his brain is doing isn't important at the moment. Maybe it is better that I can't monitor the empty sky in his fractal of the universe? If I could, it would only be another source of worry, for I know that artificial ventilation affects the brain.

At the National Institutes of Health in Maryland, a group of scientists studying the genes involved in psychiatric disorders report they exclude ventilated patients from post-mortem studies because ventilation allows the brain to become acidic very quickly after death. An acid environment accelerates the destruction of cells, proteins and RNA, and scientists need cerebral components to be in optimal condition for their studies. Can the brain become more acidic while you're still alive? The acidity (or pH) of a living brain isn't easy to measure. It takes energy to maintain the pH of any tissue and the brain is the greediest organ of the body. A healthy, active adult brain uses up to 40 percent of our energy. Consumption continues even when we sleep. But in my father's unconscious state, the brain can manage for several days on less glucose than is contained in a small piece of chocolate no bigger than a Hershey's chocolate kiss.

The further injection of propofol makes no difference to the output from any monitor.

Shona reverently sits down next to me.

'What percentage of people recover from being this ill?' I ask.

'Oh, it's hard to say. Every case is different.'

How's my father different? Is different good or bad? 'What do you think? You must have seen a lot of patients in here.'

The kindly nurse shrugs her uniformed shoulders. 'You can never tell who will survive. I just don't try to guess. People sicker than your father have made a full recovery.'

'Do you think he can? Really?'

'The doctors will soon be here on their rounds. Maybe they'll be able to give you some more information. If you wait in the relatives' room we'll let you know when it's OK to come back in,' says Shona, as she tries to tuck the escaping clump of thick blond hair back into her J-cloth cap.

She has been here as long as I have. No, longer. She has been by my father's bed much of the night. With her peachy skin and sharp blue eyes, she is a presence he would have appreciated. I would rather have been here too, to know how often he gets another few milligrams of dopamine, to watch his systolic pressure edge back over 90, to see how far his temperature deviates from 98.6 degrees.

'Yes, yes, of course. It's time I got a cup of coffee, anyway.'

Evicted into the relatives' room I can at least update my mother. But with what? He is unconscious. There is nothing new to know.

'How is he?' She grabs for information. 'The doctor I spoke to earlier says he's still critical but stable.'

I have come to hate that overused phrase. The simple triad of words at first seems comforting, calming, as though all is under control. But it is a medical deception. 'Critical but stable' implies total and brutal intervention. It isn't possible to be critical but stable at the scene of an accident—you would be dead. It really means being kept alive by artificial means until able to support one's own essential body functions again.

In my father's case, as in most, it is a busy business keeping him critical but stable, and the ICU team have worked hard all night just to be able to say there is no change. Normally we control our physiology without thought; it runs on autopilot. We do not will more adrenaline to be released to make our hearts beat faster, nor tell our kidneys to flush out toxic metabolites. But when the body's physiology is pushed outside its normal range, what was automatic becomes deliberate as the consultants in ICU take manual control. To keep Billy stable the list of drugs, tests and interventions is long and complex; two different natural hormones, dopamine and nor-adrenaline, which constrict blood vessels, are constantly infused into his veins to stabilize his blood pressure. Saline, liters of it, is infused to help keep his blood volume near normal. Then there is verapamil, a drug to synchronize the beating of his heart, which has begun to lose its rhythm, as often happens when close to death. The infection is being treated with antibiotics and there are additional effects that have to be treated too. Blood cells, fragmented by the marauding bacteria, are replaced with several pints of O positive from generous donors. His temperature, if left untreated, rises to 107.6 degrees Fahrenheit—dangerously high—and so there are acetominophen-type drugs to keep that under control.

'He's just the same, Mum. The doctors are doing their rounds so we might know more after that. I'll go downstairs and get us a cappuccino,' I say as if a small treat, something better than an ordinary coffee, could make a difference.

There is a line at the little kiosk; the seating area is busy. There are men with sallow features reading newspapers in their pajamas. They sit in wheelchairs with drips attached and nestle up to tables with healthier-looking occupants. Oh, to be them! The chaplain walks past and nods to a couple of large, grey-haired, Italian-grandmother types dressed in black. I look the other way, busy myself with collecting up corrugated cups and napkins and putting away my small change. I head back upstairs, past the flower shop with its huge tubs of *Lilium longiflorum* and stargazers appealing to passers-by, asking to be bought. I would buy some, but these are not the ones I want and, anyway, I tell myself, they would never be allowed in Intensive Care.

My horizons close in like a thick winter fog. A few days ago I had thoughts and opinions. Things mattered then. Looking at experimental data from the lab and making decisions about which compound to test are meaningless to me now. My attention is completely focused on one immobile grey body. The same information keeps rolling round on my computer screen—no mental space for anything else. Anyway, it isn't possible to run two parallel streams of consciousness. We can't hold a conversation on the phone and concurrently read a book. We can't do two things that require conscious attention. Most of us, unlike the legendary Gerald Ford, can walk and chew gum at the same time, but that's only because both actions are automatic. If we had to think about them both, we'd struggle.

This observation has opened the way for scientists to consider what the neural correlates of consciousness might be. When we look at something ambiguous like the classic picture of a candlestick formed from two heads facing each other we are conscious of either the candlestick or the faces. Even though we view one constant picture, we never see both images together. With one picture and two alternative conscious views it should be possible to find neurons that correlate with conscious perception. Professor Nikos Logothetis, working in the United States, has done just that. Using monkeys trained to indicate which form of a picture they see, he has been able to identify neurons in the visual part of the monkey's brain whose firing pattern changes along with his or her perception. It is a clue, a property of the brain, like neurons firing in bursts forty times a second, that neuroscientists strongly suspect has something to do with consciousness.

We watch for any sign, any data, that might indicate change in my father's condition. I scan the instruments around him, looking for some clue. I go for coffee. I go home. I come back. The path from the relatives' room to the chair beside my father's bed to any of the coffee places on the hospital concourse and then home again to the short-lived respite of my children's normal routine becomes automatic; I move from one to the other without thinking where I'm going or why. These are the only places I find myself, and the repetitive journeys are no more than pacing up and down on a grand scale. Except that the chair beside my father's bed is the site of least anxiety. Time spent anywhere else is time spent wondering what's happening here. At home, the phone can't be left unattended. We install two extra, one in the bedroom and one in the kitchen, just to be sure I can hear it, should it ring, day or night. I program the ICU into the memory of both and, if that weren't enough, I finally succumb and buy a cell phone.

Hours stretch into days. Two days and three nights disappear into the vacuum created by hope. At my mother's suggestion we regroup at my parents' house. As one, my brothers and I slump onto the sofa. I rest my feet on the huge marble coffee table that my parents brought back from a holiday in Mexico. I wouldn't dare do it if he were here. As we try to make something out of the days' scant events my mother appears in the doorway with a bottle of Billy's best champagne, grasping it by the neck with a tea towel. She ignores a roomful of unspoken disapproval, pops the cork with the inexperience of a woman who has always relied on her stronger half for such duties, and pours the contents into five of their best Waterford crystal flutes. 'To Billy,' she says, convinced that, now he has definitely survived more than the critical first forty-eight hours, recovery is his by right. I want to join in, to raise my glass. I want to drink, but the liquid turns sour on my tongue. There is something jarring about it, like laughing in church. The context is all wrong. My father has done well to survive a couple of days (if, in fact, he had any part in it) but isn't this all ridiculously premature?

And what would he say? What if he came round to discover that, while he lay unconscious with the Intensive Care team fighting for his life, his beloved family were helping themselves to his best Laurent Perrier '96?

***

. . . and then there crept
A little noiseless noise among the leaves,
Born of the very sigh that silence heaves.
From 'I stood tip-toe upon a little hill' by John Keats

The next day, the tiniest signs of improvement appear on the ICU charts. A little less noradrenaline is needed to maintain his blood pressure. His chest remains clear of infection and the swelling around his face is going down. We have been so absorbed by whether he will survive the first two days that we haven't thought about what might come next.

'Well, your husband has kept us very busy,' says Dr. Park to my mother at our second meeting. Now that my brother Stuart has joined us, the consultant's minuscule office seems even more crowded than the first time, despite Andrew's absence—after forty minutes he is still driving around the hospital complex desperately trying to find somewhere to park.

'On Saturday your husband was extremely ill, but it seems as if the antibiotics have indeed begun to bite and—I say this cautiously—his condition is slightly improved.' This time, the consultant's carefully measured words are an attempt to keep optimism rather than pessimism in check. 'Our plan is to keep the infection under control, then to wean him off the drugs controlling his blood pressure—a process that could take a few days. After that, if things go well and he does not develop any complications, we might lighten the level of sedation.'

'And the prognosis?' I say, using a phrase I picked up from one of the medical textbooks in my office. What I really want to ask is 'Will

he live?' But no one ever uses that phrase. It would allow the possible
answer 'No'. Worried relatives need a form of words that is vague
enough for a range of interpretations, vague enough to allow an unscientific, but sensitive, amount of fudging.

'Still possible to make a full recovery,' the good man replies. Even
though Dr. Park stresses the word 'possible', it's more than I expect,
more than I dared hope. He says something further about potential
long-term damage to his heart and, more acutely, to his kidneys, but
I am more worried about something else.

'Will he be able to see?' I can accept that my father could be on
dialysis for the rest of his life, but I can't bear to think of him as a
blind old man sitting by the window with a tartan blanket over his
knees.

Dr. Park doesn't anticipate a problem but 'there are always sur-
prises in Intensive Care medicine'. Not the least of which is what
comes next. 'Your husband is a very interesting case,' he says with
the reassuring manner of someone who has seen many. In his considered opinion, the unusual way my father's illness presented and its
startling rate of deterioration afford the sort of information that other
specialists could learn from. He asks for permission to include his
patient's case history in a lecture he will be giving at an international
meeting. For this, it will be helpful to have some photographs to
illustrate how the illness developed.

My mother agrees without hesitation. There is pride in knowing
that her only partner in life isn't bog-standard microbial morbidity.
He is special, as a photograph of the scabby black crescent under his
left eye will illustrate. Shame, though, that they couldn't have taken
it a few days ago when the lesion flared like Apollo 13 all the way
back to his ear, when his neck was twice its current size and his lips
looked more like Mick Jagger's than his own.

Slowly, gradually, the amount of dopamine and noradrenaline is
reduced further. It should only be cause for optimism, but each small
change brings some new worry. I continue in a state of anxious un-

certainty. And, as Bertrand Russell said, 'Uncertainty . . . is painful, but must be endured if we wish to live without the support of a comforting fairy tale.'

I am learning fast that uncertainty is the default position in Intensive Care medicine. I didn't expect my father's kidneys would stop working. In the hierarchy of necessary organs, the brain, I knew, would be the last to go, but who would have thought that the kidneys would get voted off so soon? I didn't anticipate the further interventions, the femoral line to connect him up for dialysis, the naso-gastric tube to feed him with a synthetic mix of nutrients. I didn't think that an unconscious person could get diarrhea.

I've never had so much time to think. Last week, the man in front of me was part of the landscape; background, familiar, the grass in the field, the clouds in the sky. I thought I knew the environment: the roads, the trees, the shape of things, how to get from A to B. My father's existence was a baseline assumption, one of those facts so obvious that it requires no examination. Now I pore over his destiny from every angle. I turn it inside out, hold it upside down and shake it. I try to understand it, to know it, and in regular acts of sheer futility I try to predict it.

Oh God, let him survive. No, I don't mean God—not like that: not like praying or swearing. Prayer is what children do, with hands together and eyes closed. My parents took us to the local Presbyterian church each Sunday when we were small. Prayer is hoping against the odds. It is kneeling by the side of your bed in the run-up to Christmas, wishing for a bike when you know that four bikes are well beyond your parents' means. It is, as C. S. Lewis wrote, posting letters to an unknown address. I've never been one to pray and it wouldn't be right to start now, just in case, if God exists, he doesn't appreciate hypocrisy.

It's OK for the rest of my family, though. My mother and my older brother, Stuart, are regulars in church. Strange, that, for a physicist, wouldn't you think? Even Andrew tries to pray, but when he knocks on the door of Thaxted church he finds it locked for its own protection. My husband and his Catholic family are the true professionals at praying. They pray for God's attention, for His will to be done. My mother-in-law willingly goes to church every day, where she lights candles for my father and has his name included in Sunday morning mass. Ripples of prayer spread out from Cambridge and from my husband's hometown of Belfast like standing waves. Within a week, people we've never met send mass cards with pictures of Jesus or the Virgin Mary on the front. The alien envelopes flutter through the mailbox as unexpectedly as Christmas cards in June.

There are, I discover, a few published studies which conclude that praying is worth the effort. Far from an avalanche of proof, but there, for the record, in the *British Medical Journal* is an article entitled 'Effects of remote retroactive intercessory prayer on outcomes in patients with bloodstream infection: randomized controlled trial'. What could be more relevant? The results apparently showed that patients whose families prayed for them had a shorter stay in the hospital and their fevers lasted a shorter time. A beneficial effect on coronary patients has been reported too. In fairness, both studies have created a furor of controversy. A quirk of statistics, a freak of medicine, or just bad science? Who knows! Observation without explanation is most unsatisfactory. Unless there is something linking cause and effect, the scientist's default is disbelief.

Anyway, religious sentiment can be artificially generated in the brain. The Canadian psychologist Michael Persinger and his colleagues claim to have found the part of the brain responsible for religious experiences. The so-called 'God module' is more commonly known as the temporal lobe. It lies just behind the temples and, spookily, it's the spot that my Religious Studies teacher, Miss Cave, used to poke with fervor when she instructed the class to 'Think, girls!'

Evidence for the structure's involvement in religious experience is twofold. First, sufferers of temporal lobe epilepsy can have deep religious experiences during and after seizures, when they see profound meaning in previously inconsequential things. Such feelings range from a joyous sense of oneness with the universe to religious fanaticism. Indeed, academics have proposed that the angelic visions experienced by Joan of Arc were due to temporal lobe epilepsy. St. Paul's vision on the road to Damascus and, more recently, Ellen White's spiritual initiation of the Seventh-Day Adventist church might also have had similar origins. Second, Dr. Persinger has been able to replicate similar sensations in a few normal volunteers by artificially stimulating the temporal lobe using a transcranial magnetic stimulator (which fires powerful magnetic pulses onto a defined part of the brain). Some volunteers, like the epileptics, experienced strong emotions and described 'a sensed presence'. Like many scientists seeking to understand their observations, Dr. Persinger suggests that there may be an evolutionary advantage to religious thought. When mankind has lived through disaster, famine and torturous times, activation of the temporal lobe, whether a consequence of mild, transient epilepsy or divine communication, has allowed the human species to persist against the odds and survive.

Shona continues with her daily routine. She flushes out the lines into my father's arm and groin to stop clots from developing, cleans his airways, brushes his teeth and combs his hair. She hums as she goes. If more volume were acceptable she might sing. I bet Shona has a good singing voice, with that husky tone and those rich, rolled Rs. She probably does karaoke. Throughout, she treats her unconscious ward with the respect a fully responsive person would deserve, sometimes breaking into a running commentary, explaining to my father what she's doing and why she's doing it. When she talks to me she uses a different voice; concern, clarity and empathy are uppermost. The information is definitely addressed to him, reminding me of the way a mother keeps a dialogue going with a young toddler, asking lots of reflective questions, pausing for replies and then filling in the answers herself.

'I'll just clean around your mouth for you, Bill. That will feel better, won't it? Yes. I am sorry about chipping your front tooth yesterday. The metal end of this suction device is much more solid than I realized. I expect the dentist will be able to patch that up later, won't he? Yes. It looks to me like it's capped, anyway. Well, that's that wee job done.'

As she tucks his intravenous line back under the sheet she says, 'Bill, can you hear me?' in the same way she has done umpteen times before: matter-of-fact, unexpectant.

'You just never know if they can hear you. You hope and try and try, but you never know,' she later recounts.

There is no warning: no twitch or quiver, no change in tempo. No alarm sounds from the monitors to signal that this time would be different from any other, but it is.

He opens his eyes.

For Shona, it is the highlight of her day, her week, who knows, maybe her year. Goodness knows, the pay's not much. And when she utters the words, 'Your father has regained consciousness. You can go in and see him,' with a smile the size of a lottery-winner's, I feel my heart will burst.

It is pounding as I race through the door, for once ignoring the sink and taps. What am I to expect? What if he doesn't know me? What if he doesn't know himself? I wipe my eyes on the back of my sleeve. What if, what if . . . .

And when I see him, it is like seeing my newborn children for the first time, only better. The inanimate grey body is gone. This is my father, my dad. He is mine. The tubes sticking out of him, the monitors, the silence, all fade away. Those open eyes change everything. Those puffy red crevices contain everything I need to know.

The corners of his eyes smile.

Behind the glassy surface is a forty-year-old bond as clear and edgy as the Waterford flutes we have raised in his honor. I can tell from the way he looks at me that he knows who I am! Without doubt, he knows. Yes, my father is back. Now, the taste of champagne arrives. This is the moment of victory. Synchronous firing, feeling and meaning are restored. The stars are shining again. He is a conscious mortal, like me. *He is back.*

I stare for a few moments. He is surely conscious, but how conscious? Conscious enough to understand me, to acknowledge me? Say something. For goodness' sake, speak to him.

'Dad, we're all so very proud of you,' I muster with a catch in my throat. Hardly a fitting speech for a prodigal father and, frankly, a little pathetic given the time I've had to think about it, but it is the best I can come up with. I feel the embryonic smile on my own face and recognize it as one I have seen in the Intensive Care Unit before. There was a girl with meningitis who spent no more than a day in the bed opposite my father. She was probably less than sixteen and very pretty, even with her strawberry-blond hair matted against the side of her head and a feeding tube burrowed into her willowy neck. She looked too weak to move and her father never left her side. There was the same slight upturn to his lips: the beginnings of a smile that lacked confidence, that withered in the bud. Talking was too much of an effort so her father just sat as close as he could, watching her with an intensity that the nearest and dearest of well people don't have time for.

Whatever the undiscovered mechanism that generates consciousness is, it is back in operation. No scientist is close to understanding how such a marvelous property exists. Analogies, descriptions and theories are all we have to explore how feeling and meaning emerge from neuronal activity. What binds neuronal function together to generate consciousness? There are as many theories as there are researchers. Dr. Rodolfo Llinas from the New York University Medical Center believes the seat of consciousness could be anatomical; that it originates from the thalamus, the brain's central core, rather like the fleshy green core of a dandelion through which all the petals are connected to the stalk. The hub extends to all regions of the brain and might somehow bind together different perceptions and sensations to provide the total experience that we understand as consciousness.

Others, including Susan Greenfield, Professor of Pharmacology at the University of Oxford, believe there is no such dedicated structure; rather, that binding is temporal; that great assemblies of neurons transiently link together to produce consciousness. On arising, competing thoughts spread like waves from raindrops on water; the largest ripple recruits the largest wave and dictates the content of our conscious stream.

Yet others propose that neuronal activity is bound by chemistry, rather than by time or place. Could acetylcholine be the neurochemical transmitter that integrates different components of the brain together? Or, as Professor Stuart Hameroff and Roger Penrose propose, is consciousness a property that emerges, ghost-like, from deep within the skeleton of nerve cells? Their theory belongs in the realm of physics rather than of neuroscience and I interpret their view of consciousness as being more like (although real experts would tell me that, technically, it couldn't be less like) a hologram, generated with fiber-optic light energy.

Dandelion, ripples on a pond, chemical cocktail or hologram: these are all concepts and all unproven. We simply do not know how a tangled mass of cells inside our heads forms a self-organizing system, hard-wired and yet still malleable, capable of generating consciousness powered by oxygen and glucose—nothing more than the same fundamental fuels other organs need. Nevertheless, the idea that a higher level of function emerges from a mass of simpler components by following a few straightforward rules is not unique to the brain. The ability of a colony of ants to structure an entire insect city with different sections for processing food, producing and feeding progeny, and with even a rubbish tip and crematorium placed outside the city limits at the farthest possible distance from each other, would serve as a good example. There is no ant capable of such deliberate design, no super-intelligent insect presiding over the others, and yet the blueprint repeatedly emerges in ant colonies the world over. There are increasing examples of complexity arising from a few simple rules.

How conscious is my father? The question is a good one. As the hours pass I can discern some fuzziness in his level of consciousness. He is awake and aware and, even though he can't speak, I can tell he is thinking, but the expressions on his face indicate that consciousness is not complete.

Consciousness is like daylight. It's not an all-or-none phenomenon. My father's state of engagement with the world around him doesn't have the brightness of a summer's day. It's more like the glow in the east before dawn. He has cycles of wakefulness and sleep, and on those criteria he passes the most primitive level of human consciousness. The next level would include registering what is going on in the surrounding world and what's happening in his own body.

Being aware of internal cues—hunger, thirst, the need to empty bladder or bowels—requires a greater level of consciousness. He later tells me that the first thing he was aware of was the desperate coldness of his feet. They are unbelievably cold, past any scale of cold in his personal history, past throbbing-cold and well into painfully cold. If someone touches his toes they might snap. It should have been no surprise. When I look at them, the tips are black and frostbitten. Keeping my father's brain and heart supplied with blood has been at the expense of his hands and feet; starved of nutrients, the most peripheral cells have died. Underneath his long horny toenails the flesh is flaking off; not as small flakes of pink skin like the peeling sunburn he once experienced after falling asleep on a camping trip with his feet outside the tent, but as strands of blistering grey. Even with most normal physiological prompts overridden by the wonders of Intensive Care medicine, it's clear from the sensation he experiences and the way his eyes follow people moving in and out of the ward that my father mostly achieves the second level of consciousness, too.

After that it's questionable. I wish I had been there the moment he returned to our world. I wish I'd been the first person he set eyes on. Instead it was Shona who reassuringly bent over his inanimate face. 'You are in good hands, Bill. You have been very ill and we are making you better,' she said. I wonder if he understood what she was saying. Did he think he was dreaming? And even now there is a tidal emptiness in his eyes, sometimes acknowledging what is said but not fully comprehending. My father can create a whole language out of the smallest nuance of facial expression. He manages a raised eyebrow for 'Tell me more' and a slightly furrowed brow for 'I don't understand'. Unable to nod, he frequently responds to simple statements or questions with a thumbs-up. Mum's on her way. I'll be back in a minute. Is everything all right? All these receive the thumbs-up.

It's quite amazing how much we can communicate without speech. When Professor Sir John Hale, the art historian and author, suffered a stroke he became aphasic. The only words he had left were 'da woahs'. Yet with only nonsense for language, his wife Sheila still felt they were able to communicate. She learned to read her husband's wishes and thoughts from the way he waved his hands, coupled with the nuance of his facial expressions. Complex conversation between my father and me is more of a challenge than it would be for John and Sheila Hale. I can't read his face; the shape has changed and his various grimaces and facial contortions are unfamiliar to me. The bags around his eyes are gone, the normal lines aren't there and his mouth is held in a permanent un-interpretable 'O'.

It doesn't seem appropriate to launch into what can only be a one-sided conversation. Why go into details of the five days that he has missed? If he ever asks me I promise myself I'll reply honestly (except for the bit about wanting him to die, obviously), but I won't say anything now. And would he understand anyway? Higher levels of consciousness require not only registering what's going on but reacting appropriately, transferring information into feelings, thinking about what's happening and making predictions or decisions based on it. And at that level my father still seems to be groping around in pitch darkness.

Consciousness is variable. Newton's understanding that gravity varies with location and condition offered new insights in physics and planetary science. The variability of Earth's species such as the subtleties of a family of birds' beaks led Charles Darwin toward the theory of evolution. Our understanding that life varies so hugely in form stretches us to define its limits. If, as Bernard Baars of the Neurosciences Institute in San Diego suggests, scientists start considering consciousness as variable, a property that becomes more elaborate throughout evolution or human development, it will lead us to a better understanding of its true origin.

My father's incomplete engagement with the world includes failing to register the relevance of seeing his eldest son by his bedside. My goatee-bearded sibling returned to the United Kingdom on the first cheap flight he could find. A two-hour stopover at Schiphol meant fifteen hours traveling in the air with no way of knowing the extent of the family crisis waiting to meet him. Billy offers him no more or less than the same wide-eyed acknowledgment he offered me.

My father is aware that he can't move but he is too weary to move anyway and he's pretty comfortable where he is, just staring at the ceiling. I notice that he stares at the ceiling a lot—in fact, this takes up most of his waking hours. The white, or nearly white, ceiling tiles have the texture of a hastily plastered wall, part smooth and part finely graveled, in a random pattern. He later explains that there had to be a grander design. In his mind, he was sure that if the squares were rearranged they would make one complete coherent pattern, like one of those children's puzzles with fifteen interlocking squares that must be slid around endlessly in a hugely frustrating attempt to recover the original design. If he just concentrated hard enough he'd be able to move them around and reveal the bigger picture. He tried willing them to move—first individually, then en masse—but to no effect.

As his daylight dawns the comings and goings in the ward around him are a performance played out behind a gauze curtain. He isn't bothered by anything. The priest administers the last rites to the liver-transplant patient in the adjacent bed. I look to Billy for some reaction—maybe fear, maybe empathy—but his countenance doesn't change. He knows that man will die. He knows that he is in the same room, with the same nurses, the same doctors, the same paraphernalia around his bed, and yet somehow he doesn't see himself as part of the same scenario. He knows he is desperately ill and at the same time he is blissfully secure and confident of his own existence. He can do anything, even will ceiling tiles to change place. The unspoken laws of life and death simply don't apply. Such a lack of reaction is reminiscent of a condition described by the nineteenth-century French neurologist Jean-Martin Charcot as *la belle indifférence*, the contentedness that accompanies mental illness, in which the sufferer seems serenely unperturbed—an effect that is permanent and probably developmental in the mentally impaired but, in my father's case, it's to be hoped, only transient.

On Saturday, after a week in Intensive Care, my father is weaned onto an oxygen mask, his lungs strong enough to inhale for himself, provided the gas isn't diluted with any unnecessary elements. The great pleasure in removing the ventilator is that it allows him to speak again. For my father, it is an obvious relief to be able to communicate by more than signs. He draws me up close and whispers very slowly. The sound of his weak voice is distorted through the oxygen mask but the Scottish accent—the Sean Connery sort of Scottish accent, rather than the Billy Connolly one—is unmistakable, reassuringly his.

As if to confirm how little he has taken in, he asks: 'What happened, Ruthie? I know you'll tell me what's going on.'

Confident that he has heard it all before, I explain about the infection, that his kidneys have stopped working. I try to choose my words with greater care than Dr. Park chose his. When my father was unconscious I couldn't say the wrong thing or cause him pain.

'You can still make a full recovery,' I blurt out, and am immediately aware that the words don't sound as reassuring as I meant them to be. I must be much more careful. 'What do you remember about getting ill?' I try instead.

'Nothing! All I know is waking up here with really cold feet. Where's your mum?'

'She's outside. I'll go and get her.'

From now on, every conversation is a matter of judgment because my father can replay my words in his mind for hours after I've gone. I know he does because he asks later for clarification of something I have said. What I don't know, I find out before the next visit. I am the biologist in the family. I can give him something the other members can't, taking care to repackage the information in crisp, fresh reassurance.

There is an other-worldliness in being so very ill. Maybe it's the drugs or maybe it's a rebound effect of five days of cerebral inactivity, but my father maintains that serene, unreal sense of security, as if no harm can come to him.

I discuss my father's altered state of consciousness with Jeffrey Gray, the retired Professor of Psychology from the Institute of Psychiatry in London. Does it help to think of consciousness as what you see on a computer screen, the objects illuminated by a theater spotlight or those flashing lights in Times Square? Jeffrey raises a different question. 'Why are we conscious?' he says. I have only wondered 'what' and 'how'.

What he says—or perhaps, more exactly, what I understand him to say, for Jeffrey is an exacting sort of person—is quite a revelation. In his view, consciousness is very much an 'after-the-fact' sort of event. It does not dictate what we do, but monitors what we have already done. Jeffrey believes that we have evolved a constant and automatic monitoring system to survey what is going on around us without paying direct attention to it. Our brains subconsciously take in information and process it in parallel. When something noteworthy happens, then it is moved up into our single conscious stream, making us actively aware so we can think, plan and react. As with the Times Square newsboard, we are aware of what is happening when it is illuminated on the board. By the time news appears in lights it has already taken place.

I imagine that most people, like me, believe that the sequence of events would be that we first think about something, then we decide what to do and finally we do it, we execute the task. Recent evidence says this is not so. We act before we are consciously aware, not the other way round. Even for a stimulus as strong as pain we react automatically before we're conscious of any action. When we burn ourselves on something hot, we withdraw our hand automatically in response to temperature sensors in the skin. Awareness of the pain follows up to half a second later. In the 1970s, the neurophysiologist Benjamin Libet showed that the brain has to be stimulated for up to half a second before it registers a conscious experience.

Fewer of our actions are under conscious control than one might think. In fact, most are on autopilot and we live up to half a second out of synchronization. I remember that as a student I came out of the bar one evening to see two children screaming on the balcony of a burning flat. I was rooted to the spot. I couldn't move. In the same fraction of a second another person came upon the scene, who, without hesitating, stopping to think or even to look round, pulled his friend along to help him bash down the door, ran up the stairs, dragged the children off the balcony and brought them out. My instinct was to freeze, his to move. I would like to say that I reasoned that I couldn't break the door down, or that I thought it was too dangerous to go inside, but it wasn't a reasoned decision. In a similar situation the same thing would probably happen again.

Jeffrey's favorite example to explain why consciousness takes place after, rather than before, an action is that of a tennis player. When Tim Henman serves at 100 mph, there is not enough time for his opponent to see it, transmit the signal to the visual cortex, then relay it to the centers controlling the movement of his body so that he can return it. Neuronal communication is slower than a fast service. Returning serve is automatic, not conscious, and the more things that can be carried out automatically, the better.

Consciousness, according to Jeffrey Gray, is a late error detection system, the brain's scrutineer. Our brains are constantly comparing what happened this time with what has happened before. On a micro-scale we subconsciously check whether things add up, whether everything in the picture fits. Hands that grip, eyes that see and a memory that keeps a retrievable record of important things are, arguably, evolution's heritage. Evolution would surely favor the species that runs on autopilot, leaving consciousness to scan for anything out of the ordinary, even though to see our feelings and perceptions as no more than a rolling snapshot of the neural processes that have already happened would seem to depersonalize humanity; to make us barely superior to the most sophisticated replicants in Ridley Scott's science fiction film *Blade Runner*.

You have to be the most detached scientist to accept that our thoughts, emotions and the meaning we give to things might be no more than the by-products of billions of well-orchestrated electrically active cells. We want to believe that, somehow, we are more than that.

The pieces of the jigsaw will one day come together. Using observation and experiment, scientists must explore what it is to be human. Most of us can only hope to add something small, to make our contribution as confidently as the data will allow. Dispassionate, logical rigor is required to advance science. Think Darwin or Newton or Einstein. To understand consciousness we will have to put emotion aside.

Yet something is lost in reducing my father's condition to a predictable pattern of neuronal activity. What joy is there to be gained from trying to describe his return to consciousness in a handful of equations on a piece of A4? The facts can't be reconciled—no matter how hard I try. How can consciousness, the elusive 'landscape of our sovereign will' that Auden so beautifully described, span all the complex thoughts and feelings I have for the man who shaped my life and, at the same time, be no more than assemblies of cells spewing out electrical discharges forty times a second?

4

The Balanced Brain

*The management of a balance of power is a permanent
undertaking, not an exertion that has a foreseeable end.*
Henry Kissinger in *The White House Years*, 1979

WITH HIS SEPTICEMIA UNDER CONTROL, MY FATHER IS ABLE TO BREATHE
under his own power. He is medically defined as stable and, as such,
qualifies for transfer out of Intensive Care. More importantly, perhaps, the ward is full and his bed is urgently needed by other more
deserving cases. He has graduated from critical but stable, *magna
cum laude*. We wash up the pink-bow cups in the relatives' room,
throw away yesterday's newspapers and say goodbye to Isobel, whose
husband continues to stack up complications like a house of cards.
She hugs my mother, just as she did the other relatives that spent a
few days in her company, taking a little vicarious pleasure each time
someone else makes it out.

My father is one of the lucky escapees. I could dance down to
the lobby and eat an entire egg-and-crispy-bacon baguette. I could
down that glass of champagne in one gross swallow. I could kiss Dr.
Park, Dr. Uli, Shona, the entire ICU team.

Life is very different back on ward C5. The calm order of a high-
tech environment is traded for a hot, sweaty, laissez-faire atmosphere
in which my father is the least independent patient. Surrounded by
the failed liver-transplant case and the comatose meningitis victim,
he looked darned good. But now his peers eat off plates with their
own forks and wear slippers to visit the bathroom at the end of the

corridor. Billy is physically impoverished by comparison. He can barely move unaided and, as for visiting the conveniences, putting a pair of slippers on his feet would be like trying to squeeze a balloon into an egg-cup, and to what purpose? The damp bag at the side of his bed is all the toilet facilities he needs.

And how does he react to his new surroundings? Were it not that his limbs have become functionless slabs of flesh, I would have said that my father takes it all in his stride. He seems pathologically unconcerned. Mentally, he can do anything he wants. So, when he wants a drink he reaches for the jug of water on his bedside table. Lacking the strength to stretch that far or the presence of mind to press his buzzer for assistance, he tips himself out of bed straight onto the floor, in so doing ripping out every drip and tube; and he lies there, laughing loudly, as if it were funny, until four able-bodied staff can lift him back into bed.

'Typical Billy,' says my mother, smiling. 'No patience!'

His reactions are inappropriate and scary. It will take some time to get his head round everything that's changed, I reassure myself. The brain is an intricate system of control, like a town that has evolved a set of smoothly running operations. When everything stops it's going to take time to get it working seamlessly again. After five days' total shutdown in Intensive Care and five more in a state of serious drug-induced slow-down, we're not talking about a bit of a social disturbance. We're talking mental anarchy akin to the looting of Baghdad. It will take time for the highly complex meshwork of neurons to resynchronize.

The architectural arrangement of neurons in the brain has a consistent but untidy pattern. There is a line of symmetry right down the middle, but the brain lacks a discernible geometric design although it does have multiple regions, nodes, tracts and bundles. A cross-section of one hemisphere looks more like the randomly evolved street plan of an ancient city like London than the predictable grid plan of Manhattan. The Spanish forefathers of neuroanatomy Santiago Ramon y Cajal and Camillo Golgi were the first to cut a brain into wafer-thin slices and reveal the richness of its detailed structure. The apprentice shoemaker and his friend stained sections with dyes and through a high-powered microscope they could see the tree-like shape of a neuron with its dense core (the cell body containing the nucleus and much of the cell's necessary biochemical machinery forming the central crown of the tree) from which multiple dendrites fan out like branches. A slender trunk (the axon) leads down to the smaller roots of the tree (the neuronal terminals). Neurons communicate by passing electrical signals from branches to root and then on from the roots of one tree to the branches of the next. Cajal and Golgi's pioneering work, which showed that each brain region contains neurons of many different shapes and types, won them a Nobel Prize in 1906. Take the hippocampus, for example, the region that the famous amnesiac HM lacks. Under a microscope it looks like three deciduous forests in winter, stacked one on top of each other and curled round to make the characteristic sea horse shape.

It's not surprising that rebooting my father's brain is taking a while. The brain contains about 100 billion neurons and they're much more dynamic and active than their structure might suggest. Each neuron can 'talk to' thousands of neighbors through connections called synapses. These occur between the terminals— swollen root structures of one neuron and the dendritic branches of the next. Information is carried electrically along a nerve. When it passes from one neuron to the next in the circuit, the signal is transformed from electrical to chemical. The normal brain is continuously active with information passing from neuron to neuron—from root to branch—and on in reverberating circuits—from electrical to chemical—over and over again, in the click of a finger.

When you think about the almost unimaginable number of synapses in your brain—more than the number of all the humans who have ever lived on Earth—it conjures up a communication pattern of incredible complexity, behind which order must be maintained. The brain has evolved many ways of managing communication, just as we have in the Western world. Physically neurons bundle together to form major tracts, like motorways, and then spread out into different regions like minor roads and paths. Temporally, the developed world can communicate at different speeds by phone, gossip or letter, for example. Our brains have similar options: fast neuronal circuits, modulators and hormones. In terms of language, there is heterogeneity too. While we speak in English, Spanish or Swahili, chemical communication across synapses is in glutamate, noradrenaline or serotonin.

Maybe it isn't just neurons in my father's brain that have to be given time to recover. Intermingled with them and equally gigantic in number are the glial cells. Scientists used to think that glia were merely building material, the mortar that held neurons together. We now know that they do much more than that: they provide a continuous supply of ready-to-hand information so that the governing neurons can work efficiently. They generate some of the chemicals that neurons need, handle any waste that can't be reused and, in ways that have yet to be worked out, they enhance neuronal development. Glia support but don't decide—they are the passive civil servants of the brain, while neurons are the government politicians.

I sit down by my father's bedside and wait, watching out through the open door of his single-occupancy room—the high-dependency bed, they call it—right by the nurses' station. When Dr. Park first said that my father was being moved to a high-dependency unit, I assumed it was something to do with drug dependency. What was he dependent on? Couldn't be opiates: he hadn't had much morphine. Hardly benzodiazepines! While he'd had a lot of the Valium-type drugs I know so much about, being dependent on them hardly constitutes need for a specialized ward. If it did, we'd have a lot more middle-aged women filling our hospital beds. I almost laughed out loud when a nurse explained. Others! He is dependent on others. Nursing care. Being looked after. 'It'll be a while before he can look after himself,' she reiterated, in response to what I can only assume was incredulity on my face.

It won't take long, I reassure myself. Billy will soon have the nurses running round in small circles. He's used to giving orders, taking charge, not being dependent. He'll be back in command in no time at all. Maybe I should warn the staff, let them know what kind of person they're going to be dealing with.

Blue-striped uniforms come and go but I wait for the registrar to appear at the desk. That must be her, the one in a white coat and dark tights (a fashion no-no for anyone but a medical professional), with sleek, bobbed hair.

'Hello, I'm Ruth McKernan. I understand you're going to be looking after my father—in the high-dependency bed,' I say, pointing in the appropriate direction. 'I'm afraid you're going to find him a bit of a challenge.'

'Oh, in what way?' she replies, tilting her head to the right to indicate that very few patients are a challenge to her.

I should have had the sense to hold back but the day's excitement drove me on. 'He is used to running his own company, you see; used to things being done as and when he wants them. He can be very demanding.' I should have stopped there but it is too late. Out pour the words: 'You must understand. He's not a man to tolerate fools.'

She puts her hands firmly into her pockets, stands firmly upright. 'Your father is still very ill. We will take things one step at a time and *we* will decide whatever is best for him.' And then, as befits her training, she soothes the intentional slap on the wrist with: 'But don't worry: give or take an infection or two, we'll have him out of here in a matter of weeks.'

I feel foolish. At the same time I feel my spirits rise. This woman knows what it will take to get a man back to health. She has the experience of seeing many pass along that conveyor belt. The important thing is her obvious assumption that he will survive and my joy is all the sweeter when constructed from facts, figures and statistics, rather than hope.

In the meantime my father has to be moved to a bigger bed with a restraining side rail, to protect him from his own enthusiasm. It didn't matter that the Intensive Care bed was barely long enough for him, because he hadn't the energy to move then. Now, whenever he presses the button to raise the head of his bed, his body slips down, leaving his feet overhanging the bottom. Tracking down a larger bed takes all day; one ward has a frame, a few more phone calls finds an inflatable mattress, and the nursing staff put the contraption together with all the confidence of novice home-owners building their first self-assembly kitchen unit. As is traditional with flat-pack furniture, after it is assembled there are a couple of small bits left over, and no one is really sure how the controls work. My brother Andrew randomly presses buttons, to the obvious disapproval of the sister, who eventually politely removes them from his clutches (although I wonder if she knows how to work it herself).

My father is much more comfortable now that his feet don't overhang and his arms don't slip off sideways. Raising and lowering the bed to watch what is going on even gives him some minimal independence. Characteristically, he goes a bit too far, puts himself almost upright and slips down immediately. It takes three or four of us to reposition him and we get well practiced in the task. Using the sheets to take his weight, we rearrange his limbs, pillows and new poly-cotton PJs with the coordinated speed and efficiency of a Formula One team carrying out a tire change.

'Dad, you have to keep the oxygen mask on. That's what they give it to you for.'

'I don't need it,' he says defiantly and waves his one mobile hand as if to dismiss me.

I look to my mother. *You tell him,* I silently plead.

'Please, Billy. It'll be easier to breathe,' she says as she puts it back over his head.

'But it's harder to speak and I . . . I can't . . . oh . . . OK.'

Helping him breathe more easily isn't the only reason for the mask; keeping a good supply of oxygen to his brain is what's important. When the oxygen level in blood falls, the first thing to go is concentration; the ability to think straight. Oxygen consumption is so critically tied to brain function that you can even tell which parts of the brain are working hardest by monitoring how much oxygen they use. When the brain's two essential nutrients, oxygen and glucose, are used up, more supplies are immediately delivered to the active parts; consumption mirrors activity.

Modern-day neuroscientists can follow brain activity in real time, in real patients, which is such an advance on the predominantly structural information that our forefathers could get from their thin, stained slices. Using PET (positron emission tomography), which monitors glucose utilization (a radioactive form is injected into the bloodstream to facilitate this), or functional magnetic resonance imaging (fMRI), which tracks the consumption of oxygen, scientists can now produce movies or videos to show the dynamic state of the brain. The best that Ramon y Cajal and Golgi had at their disposal was the scientific equivalent of a pinhole camera. However, neurons still fire thousands of times faster than an image can be collected and the smallest pixel of activity we can detect derives from thousands of neurons. Nevertheless, we can get a good general picture of the working brain by capturing sequential images from large constellations of neurons. This was most beautifully demonstrated by Professor Semir Zeki from the Institute of Cognitive Science at University College, London, who ended a lecture recently with a reconstruction of the waves of activity recorded by fMRI from volunteers who watched the opening two minutes of the film *Tomorrow Never Dies* from inside the scanner. He pointed out the different areas of the visual cortex, the dramatic activation of the auditory cortex when the music started and the effort expended by the frontal cortex trying to work out what the hell was going on. Not quite a dynamic brain video but so perfectly synchronized that the Royal Ballet would envy its choreography.

I check the oxygen line. It sounds like there is gas coming through and we can still hear what my father says if we sit close enough to smell the mask's stale plastic and see the tiny drops of condensation on its inner surface obscuring the movement of his mouth. My ear is close enough that his words, although a little muffled by the mask, can bridge the gap from lip to ear. Communication works best over very small distances.

When it comes to neuronal communication, electrical signals can't cross synapses (the gap between neurons). They are converted into chemical signals to carry information across the critical few millionths of a millimeter. Most speak just one language; they release just one type of neurochemical. Neurons that excite others release glutamate. You'll probably be surprised to know that it is the same glutamate used in Chinese food as the additive monosodium glutamate. It's the chemical that our tongues detect as a lip-smacking, tasty flavor, and receptors on neurons detect as excitation, the message to fire, to go, go, go!

The oxygen mask doesn't inhibit my father's plans. Rather the opposite: he is up for anything. A few conscious days and his cortex can barely contain its creativity. Big ideas spurt out like water from a New York fire hydrant. He and his pharmacist friend Bryce are going to overhaul the whole National Health Service. From his high-dependency bed, he can see where improvements could be made and he waffles on urgently about the appropriate staffing levels, the arrangements of beds, lighting, food, cleaning schedules, even the distribution of toilet paper. Together, they can do it. First, though, he'll go on another Caribbean cruise. He insists that my mother bring in brochures. To humor him, she arrives in her holiday attire—emerald-green slacks and a polo shirt with matching green and pink daisies (she has deliberately chosen not to wear black to the hospital since the day he was admitted)—clutching a sizeable armful of glossies from P&O, Princess and Royal Caribbean.

'Antigua, this time!' he declares, and my mother enthusiastically agrees in principle but begins to get nervous when he wants to book the May 12 departure date. In his mind these are not mere discussions to raise his spirits. These are concrete plans and he wants action.

'I'll take the big grey case,' he says to one nurse.

'Tell Catherine not to forget the binoculars,' he tells another. He announces details of the boat, the type of cabin (exterior with balcony) and the route, which have neither been agreed nor confirmed, to anyone who'll listen.

Late one night, well after midnight, he persuades the young staff nurse to phone my mother. She is to tell his son Kevin (you will have realized by now that he doesn't have a son called Kevin, although coincidentally it is the name of the owner of the local taxi firm) to collect him and take him to the docks, where his ship is ready to leave.

'The big grey case. Tell her we need the big grey case. . . . And the binoculars. Bring the binoculars,' he repeats agitatedly. Instead of checking his lines, adjusting his oxygen level and maybe considering that another small dose of Valium might be in order, she makes the call.

Who can blame her? My father is a convincing man and he can be quite uncompromising. He once negotiated the sale of some chemical technology to the Malaysian Navy by uttering only a single phrase in a business meeting that lasted seven hours. While his son stood silently by, my father said, 'The price of this technology is £25,000.' He repeated it no more than a dozen times and that was his total conversation for the day. He got his price and at the same time Andrew learned one of the more extreme forms of negotiation. He is going on a cruise. Kevin is coming to collect him. It is a fact. It is a £25,000 fact. And there is no more to be discussed.

My mother holds her head in her hands. 'What *is* the matter with him? It's so unlike him,' she despairs.

I hand her a tissue from the box on the table. 'I know. But he's not so bad. I'm sure it's only temporary,' I say as she absentmindedly winds the corner of her tissue into a corkscrew over and over again.

I'm bluffing. Out of all of us, I should be the one most able to cope. But knowing something of how the brain works doesn't make it easier to accept my father in what I hope is a temporarily addled state of mind. Psychiatrists used to prescribe deliberate coma, precipitated for example by administering an insulin overdose, to shock a mentally ill patient back to normality. Setting the brain back to zero was once a common and fairly successful treatment for depression and psychosis. But you're supposed to start out sick and wake up better, not the other way round.

Today, I stroll, rather than rush, out of work at 4 p.m. to be by his side. I don't worry when I can't find a parking place. When Billy first regained consciousness he wanted to hear what I had to say. He looked forward to my visits. He needed me. Now I drag my feet. His state of mind is deteriorating. Conversing with him is worse than talking to a stranger. At least with a stranger you don't know what to expect so anything he says is acceptable. But I have known this man for forty years and I don't know how to respond to him.

'Is that James Baxter playing the pipes?' he asks. 'Why is James Baxter playing the pipes outside my window? Is it a party? What are the celebrations for? Ruth, go and look.'

I get up and look outside. Three floors below me the scene is of an enclosed concrete patio without a soul on it. 'I can't see anything,' I say, hoping that such ambiguity will suffice.

But my father insists. He knows one of the men playing in the band. It is James Baxter, a friend from university, who was my godfather. 'Is it the drums or the pipes James is playing?'

I try to respond calmly. I try to analyze this hallucinating. Why does he imagine that James is playing the drums outside his window? Can he hear some kind of droning noise in his head, I wonder. What would be associated with a droning noise? Bagpipes? James Baxter? Perhaps his ravings have more in common with the way we fabricate a dream out of synchronous patterns of firing nerves while we sleep?

Billy continues with his questioning. His eyes grow ever more confused in disbelief that I fail to hear the same sounds he can.

The doctor arrives. I try to speak for him, to protect him, hoping that she won't notice.

But my father fires more questions at her. 'What's the celebration? Is it a party? What's going on outside?'

She peers through her tortoiseshell-rimmed glasses, smoothes her hair neatly behind her ears in preparation and steps around his questions as neatly as she might avoid a puddle on the floor. Instead she replies with some questions of her own, starting with his name. Then she asks him who I am. He gets them both right and I am as relieved as if I've taken the test myself. Next, she asks where he is. 'Canterbury,' he replies insecurely, and then he spits out a few more towns beginning with C. 'Coventry? Carlisle? Cowdenbeath?' He looks vacant and scared. C for clouded or confused would be more like it. He struggles on a little longer, clutching names from the air. He never finds Cambridge. His eyes widen further, fear seeping round the edges and welling up under the rims.

I know that sense of intense anxiety. In fact, I know anxiety from A to Y; not just because I've been working on developing better drugs for its treatment, but because I've experienced acute panic myself. It happened twenty years ago when I was taking part in a clinical trial. It was a simple experiment: just a combination of two tablets and a series of blood samples. I've never been concerned about giving blood, in fact I pride myself on the quality of my antecubital vein—a one-eyed orderly with shaking palsy could get a needle in there. But no more than ten minutes after the second drug—bang!—from nowhere I felt an overwhelming sense of disaster as though trapped in the Coliseum seconds before the wild animals were to be let in. Doom closing in, suffocating me. Nurses tried to calm me but I was past reason. I knew it was an effect of the drug, but I couldn't stop my mind from pleading: set me free. *Let me out! Make it end! Please!* No amount of wiping my hands down the sides of my jeans could dry those sweaty palms. And after fifteen minutes the anxiety, which we now know was induced by that particular combination of drugs, subsided, leaving me exhausted from the unforgettable, unavoidable terror.

I try to rationalize away Billy's odd behavior. After all, he is alive. We should be grateful. I never thought he'd make it this far and I still feel privately guilty for wishing him dead. Even if he's mentally a bit unsteady, at least physically my father is moving in the right direction. He is allowed to eat small quantities of real food now, if approved by the dietician. So I attend to the physical instead and make one of his favorite meals, fillet of fresh smoked salmon with new potatoes and lemon mayonnaise, and present the small, carefully arranged dish on his bedside tray.

His smile wanes. 'You get salmonella in eggs, don't you?' he asks suspiciously.

'Dad, I made this less than an hour ago. It only takes twenty minutes to get here from home.'

'Yes, it looks lovely, Ruthie. I'll have a potato.'

I break up the Jersey Royals and feed him a few small pieces.

'How old is the fish?'

'It's fresh, Dad. Tesco's finest, this morning.'

'I'll have another piece of potato.' He chews it slowly, as though ruminating over some deep and mysterious question.

I try him again on the fish. Surely he needs the protein?

He pauses, rolls his eyes in fear and shakes his head. After two more refusals I put the plate down and wait. He avoids looking at the food and I later silently scrape the rest into the bin.

His unrestrained zest for life is surely passing.

The brain isn't made only of excitatory neurons. They would be continually activating one another until they either drive the brain to a standstill or excite themselves to death. Excitatory neurons are matched by inhibitory ones. The chemical messenger for inhibitory neurons is GABA (which stands for γ-aminobutyric acid). As far as we know, GABA has no particularly interesting properties other than being one of the two most abundant neurotransmitters in the brain. Oh, and it is my favorite neurochemical—I've studied GABA and its receptors for a decade. GABA probably has the balance of power in the brain. Excitatory neurons would be going off like fireworks on the Fourth of July if not kept constantly suppressed by GABA's instruction to stop, stop, stop.

Inhibitory neurons restrict how far neuronal activity spreads. Clusters of inhibitory neurons also generate oscillations of electrical activity in pacemaker parts of the brain. They maintain the brain's 40Hz pulsations in the way that the soldiers marching across a rope bridge set up standing waves. Between them, glutamate and GABA are responsible for most fast neuronal communication. Movement, speech, reason; from the way the right side of my father's mouth moves first when he smiles to the selection of his thoughts and words, all are the product of a delicate interplay between glutamate- and GABA-releasing neurons.

The doctor has only just left when my father calls her back again and, in an uncharacteristically patronizing manner, asks her to have a word with the 'girls'. He is concerned about their hygiene. Can she make sure they wash their hands before coming into the room? Have they been checked over for germs? Germs—from a man with a Ph.D. in biochemistry and a cryptic-crossword vocabulary! What is he talking about? I am embarrassed. Does he think their negligence might harm him?

It would be disrespectful not to humor him, but somehow I have to find a way of telling the doctor that this is not like him. As Billy pontificates on the need for extra protective measures for the nurses—gloves, hair nets, could they wear masks?—I catch her eye. I want to make amends in the same way that my father might have apologized for some misdemeanor of mine as a child.

Puerile misdemeanors there were aplenty. Not that I was deliberately hurtful, but, well, things happen. There was the time Billy organized a surprise birthday dinner for my mother in a swanky private room at the Recorder's restaurant in Thaxted. As an impoverished Ph.D. student I was doing two part-time jobs and saving up for my first mortgage and I pretty much wore the same pair of jeans most days, washing them on Sundays whether they needed it or not. To get to the restaurant on time I had a quick shower and put on the same jeans I'd taken off; clean underwear and shirt, though—even scruffy students have standards. We washed the chocolate soufflé down with the last of the Pouilly-Fumé and were just putting on our coats when the waiter held up a pair of knickers he had found under the table. I can only presume they had been stuck in the leg of my jeans and worked their way out during the course of the evening. My brothers laughed like hyenas. Billy said nothing. He just shook his head slowly in that 'Children! What can you do with them?' sort of way.

Now it is my turn. The man who could put together a witty after-dinner speech with less than a day's effort; a man who took the honored role at the golf club Burns Night supper of reading out the poem 'To a Haggis'—in his richest, raspiest, most velvety Scottish accent; a man who could hold court in any social gathering with his endless supply of jokes aimed at the English; the skilled entrepreneur who could engage anyone in conversation, now completely lacks grace. Somehow the illness has pruned his social skills to the trunk. He is still as persuasive as ever, but with all the sensitivity of a spoiled six-year-old.

'I'm sorry. Really. He's not usually like this,' is all I can manage.

Which is worse, psychosis or anxiety? When he was psychotic at least he was happy. Those pathologically high spirits were destined not to last, and even if all that enthusiasm was only bliss accumulated from days of inactivity just waiting for the chance to be released, it was still bliss nonetheless. Anxiety hurts. Watching someone you love experience pain hurts. In fact, true empathy is a scientifically observable phenomenon. A group of scientists at London's Institute of Neurology used MRI to study the brain regions involved in pain perception. When volunteers received an electric shock, the circuit involved in pain perception (the anterior cingulate and anterior insular parts of the cortex, the brain stem and lateral cerebellum, if you're interested) visibly increased in activity. When volunteers watched their partners being given a shock, many of the same regions responded in the observers' brains too. There is scientific support for the long-held belief that we feel our loved-ones' pain.

The following day, lunch at my favorite restaurant, The Three Horseshoes in Madingley, turns into a game of shuffleboard in which I move three crisply seared baby scallops around the plate, unable to bring one to my mouth. I can eat only when sufficiently distracted. I wash fruit several times over. Yogurt and milk disgust me as I ponder the number and varieties of bacteria they contain. I wash my hands on the hour, clean my teeth more frequently than ever in my lifetime, and I brush my tongue and cheeks to rid my mouth of bacteria. I have lost seven pounds in the two weeks that my father has been in the hospital. And the shock of it all is that, despite all my knowledge, I can do nothing about it. For the first time in a decade of trying to develop better Valium-type drugs I begin to understand, really understand, how chronic anxiety feels.

At work, when I can bring myself to go in, I am restless, shivery. I can hardly sit still unless I'm fully absorbed. Full-force concentration barely keeps the image of my father compressed, out of the cortical quagmire. I probably seem cold and distant but if I allow myself to relax, to chat, to be engaged at all, anxiety becomes the default. So inquiries about my father are deflected with curt minimalism. I waste no energy on small talk. My colleagues seem to understand, or at least they're sympathetic enough not to derail my strategy for sanity. I fidget endlessly in meetings and get up to fetch cups of tea or coffee that taste too strong to drink. Wringing my hands—such a cliché, I know, but nevertheless I catch myself doing it—and scratching at my itchy skin feature high on the list of activities that fill the gaps between hospital visits. As Roger McGough wrote:

> *Don't worry, I'll do it for you*
> *Relax, I'll take the strain*
> *Anxiety is my forte*
> *I've got worry on the brain.*

Knowing all the theory doesn't really help. When I was studying for my Ph.D. I saw mental illness as a problem of brain chemistry. Generalized Anxiety Disorder is a neurochemical imbalance that we do our best to fix with medicine, and improving that medicine has been my goal for the past eight years.

I could take some Valium of course. Valium amplifies the effects of GABA; it turns up the dial on the brain's inhibitory system. Mother's little helper would bring me sleep, take the edge off my anxiety. I don't submit, though. I don't want to feel sleepy: I need to drive and I'm none too keen on withdrawal and potential dependence, either. I might have had something better—a daughter's little helper—if only we'd worked a bit harder, or maybe been a bit luckier in our research.

The hope of making a Valium-type drug without the side effects was born just across the road from my father's bedside. You can almost see the Medical Research Center building from the window. My friend Anne Stephenson (now Professor of Pharmaceutical Chemistry at London's School of Pharmacy) and her colleagues isolated the first type of GABA receptor from the brain in 1988, and worked out its DNA sequence, the genetic recipe for its construction. Since that discovery at least a dozen more types of GABA receptor have been found. Peter Seeburg and his group from the University of Heidelberg did the lion's share of the work, but there is a tiny contribution in the archives from Messrs. Whiting, McKernan et al. from the Merck Neuroscience Research Center. I can know all this detail; know the shape of a receptor a millionth of a millimeter wide, know that it is made of five units that each snakes across the neuron's membranous surface four times, know that the whole assemblage opens like a camera shutter when GABA is attached, know that Valium opens the shutter more frequently, letting in chloride, which makes the membrane of the cell less excitable. I can know the whole neuronal microcosm in all its minutia and intricacy and detail and still have no control over my own anxiety nor understand why my father's brain won't work properly.

In search of some kind of explanation, I ransack textbooks, surf the Web and sift through details of known side effects of the drugs he's taking. Relief comes in an article on a syndrome called 'ICU psychosis'. Medicinenet.com says that a third of patients who spend more than a week in Intensive Care can expect to suffer from it. The longer you're in there, the more likely it is to happen. Billy endured the bizarre existence for ten days, albeit five of them completely unconscious. Once I found the term 'ICU psychosis' I realized the phenomenon is well described in the medical literature. The cause is not clear and might not even be the same in all patients. There are physical causes: dehydration or heart failure that reduce the supply of oxygen and glucose to the brain. But there could also be other causes: the strange environment where night and day blur together and normal cues for living—meals, light, darkness, sleep—are abandoned. People who wear glasses and hearing aids are more susceptible when their aids are removed. Noisy monitors, bright lights and the sensory deprivation of being in a room without windows, away from familiar things, leave the brain without signals to reset.

Glutamate, which excites neurons, and GABA, which inhibits them, are the positive and negative of the brain; yin and yang, up

and down, plus and minus. Having too much of one or not enough
of the other leads to equivalent disruption. Anesthesia is produced
by increasing GABA neurotransmission (more profoundly than
Valium can) or by reducing glutamate transmission. The reverse, too
much glutamate or not enough GABA in the brain, leads to convul-
sions. Between these two extremes is a subtle range where modest
changes can have bizarre effects. Small doses of PCP, otherwise
known as angel dust—or any number of street terms for recreational
drugs that a woman of my age is no longer likely to come across—
block glutamate's actions at a small number of synapses and can
cause psychosis. Schizophrenia-like symptoms—hearing voices, dis-
connection from the real world and paranoia—can be caused by a
modest imbalance in glutamate neurotransmission. When the brain
has been shut down for days it might take a long time to re-establish
equilibrium. It is a reasonable understanding, I tell myself.

For my father, recovering consciousness, being weaned off the
ventilator and leaving the ICU were all temporary triumphs. Each
step brings the next hurdle into sharper focus. My problem is I know
too much. When his breathing was sustained by machine, I feared
the bacteria that might force their way into his lungs and breed up
into a tidy little infection—terminal, of course. When that failed to
materialize I started to worry about his high blood pressure. Stroke is
one of the biggest risks. Then I start thinking about the statistics that
predict his future. Forty to fifty percent of people who are admitted
to Intensive Care from a lesser ward in the hospital don't survive.
Moving in that direction usually means you're getting worse. Patients coming straight from the Accident and Emergency unit or
from the operating theater have higher chances of recovery—they're
getting better. When Billy's blood pressure returns to something close
to normal, his functionless kidneys take up the baton. He needs
dialysis to remove toxic waste products from his blood and that carries a risk of embolism. The choice now is brain or kidneys. No
contest! Brain wins every time. I would gladly parcel up his kidneys

in a bag of ice, stuff them into boxes labeled L and R, just to have his mind back.

There is no pattern to his moods. The peaks of exhilaration are sporadic and at times his anxiety levels seem normal. My father says that dialysis gives him a strange feeling, one that he doesn't particularly like but he can't quite articulate what it is that he doesn't like. Maybe it's the sensation of having his blood sucked out through the cannula in his groin, watching it gurgle through the extraneous tubes and valves of a souped-up washing machine right before his eyes, before being returned, cooler, back into his veins. Maybe it is the real and rational fear that any invasive procedure carries a small but significant risk of infection, and another infection is something he can ill afford. Or maybe it's a constant reminder of how far into the shadow of the valley he has been. Whatever the reason, he asks me to stay and chat while the nurse swabs him yellow with antiseptic and plumbs him in.

'Remember the Caesar salad that we had in Captiva Island?' I offer.

'Handmade at our table in a large wooden bowl, rubbed round with garlic. And they made the mayonnaise right before our eyes.'

The Romaine is crisper, the anchovies more piquant and the croutons crustier when served with reminiscence. We move on and mentally dine at the Ivy. Billy has been there quite a few times, though I accompanied him only once, along with my grandmother and great-aunt.

'We ate downstairs, next to that really noisy table. Do you remember, Ruth, I had to go over and ask them to lower their voices and moderate their language?'

An undisputed act of chivalry but, as my Auntie Belle had been toughened up by spending her youth playing the piano in a jazz band, I wondered at the time whose feelings were being spared.

We stuff our heads with marinated peppers, game soup and New York cheesecake; we protect ourselves with my mother's favorite dinner-party menu from the Cordon Bleu cookery course: chicken and peaches with onion pilaf followed by pineapple mousse *en surprise*, if she could get the ratafia biscuits—if not, sponge fingers would do, but they weren't nearly as good, we concur. We bury our minds in hors d'oeuvres from the special teak and glass lazy Susan that my father fills at Christmas; anything to keep those other thoughts pressed out against the meninges, away from operational parts of the brain.

Had I known that mental illness could be caused by infection or ICU psychosis but that recovery in either case was virtually guaranteed, I wouldn't have been so cavalier about Billy's kidneys. Despite the high doses of antibiotics my father takes, the registrar suspects that his psychosis might be microbial rather than psychological. Her hunch is borne out when swabs taken from the line into his femoral vein are found to be harboring one of the smartest and least appreciated bacteria of our time, MRSA. He was right to be paranoid about cleanliness. Somehow, the bacteria have got into his bloodstream. This strain is the scourge of Addenbrooke's and many other hospitals. MRSA, short for *methicillin-resistant staphylococcus aureus*, is resistant to almost all antibiotics. If streptococcus is Joan Rivers, then MRSA is the Osama bin Laden of microbes—resilient to attack, impossible to pin down and you never quite know where it's going to turn up next. We have one remaining treatment for systemic MRSA (still nothing for Osama bin Laden), a drug called vancomycin. It is reserved for use in hospitals because it can cause kidney damage, so during administration blood levels have to be carefully monitored—particularly so, in my father's case.

The bacteria in his bloodstream could be responsible for his growing mental confusion. I fear the return of septicemia. Could MRSA put him right back into Intensive Care? Fortunately, it's not the same kind of microbe. Although it causes fever and infection, it doesn't have the genes to rupture red blood cells so it is unlikely to cause septicemia. In a healthy person it's hardly likely to be detected at all, but in an already compromised individual, fever, chest infection and lethal pneumonia are still possible consequences of infection.

Although routine swabs show that my father retains the microbe on his skin and respiratory passages for many months to come, vancomycin clears the infection from his blood after two days of continuous intravenous treatment. The devils dancing in his brain quiet; euphoria and confusion dissipate. The untethered brain activity of the previous week, the delirium and psychosis, the displays of water fountains dancing to music, suddenly stop and his brain is as quiet as a duck pond. My father answers questions parsimoniously with little inflection or spirit, as though saving energy for something more important. Empty of the mental vigor the infection brought, he is more like any other recovering patient and energy is conserved for the rudiments of personal hygiene. In place of a proper wash he can now wipe his face with a damp flannel. He hasn't yet recovered the art of cleaning his teeth, which he does very slowly, in an infantile kind of way, holding the toothbrush with his thumb alongside his fingers in the grasp of a less-developed primate. He dislodges toast crumbs that cling to his chipped yellow teeth by swilling water around his mouth and spitting it into the bowl on his lap.

As I empty the contents down the sink I look at the saggy man slumped against his pillow, hand on stubble-chin. I wouldn't give much to have that rubbed across my face. Wincyette pajamas and fleecy lambswool slippers would not be out of place. A vision of freeze-framed frailty, he has aged more than ten years in less than three weeks. He hasn't had a decent shave since Shona attended to his needs. Now he looks like any average sick old man struggling through each incremental recovery. I catch myself watching him and wondering whether I didn't love him more when he was magnificently, manically mad.

The feeling doesn't last. No matter how disturbing his physical or mental state, he is still my father and it takes only those two words to transform the image of a feeble man sitting in a bedside chair into something much, much more important.

5

Stem Cells

WHEN MY FATHER WAS IN INTENSIVE CARE I COULDN'T WORK. I canceled a trip to New York to present some interesting new data on the structure of GABA-A receptors. Manuscripts in need of revision piled up on my desk like segments of spinal cord. What use is one more publication in *Molecular Pharmacology*? Will the world be a lesser place without it? On days when Billy still fails to progress—the announcement of yet another gastrointestinal infection, high white-blood-cell counts, another sleepless night—I close the door to my office and stare pointlessly at my screensaver. I mirror my father's state of mind like one of the Grand Old Duke of York's soldiers: when he was up I was up and when he was down I was down.

Following good news I rally back to work. On the day a trickle of orange fluid appears in the bag beside his bed I edit three manuscripts and call in my senior group for a brainstorming session.

'I reckon there's at least fifty ml in there, Mum.'

'Fifty ml? Looks more like a tablespoonful,' she says, staring at the liquid as though it were ancient treasure of yet to be determined value.

'Still, it's more than there was yesterday, and all of last week.'

'Maybe they'll let you have a proper drink soon, Dad.' I hand him a bowl and he takes out one small ice cube and pops it in his mouth. Who had the bright idea of giving ice cubes to those who must restrict their fluid intake? They moisten the mouth as they melt, lasting much longer than a sip of water.

'I'd really love a big glass of water. One sip isn't enough. These ice cubes are too cold for my gums and they don't really wet my lips.'

I take the Crabtree and Evelyn aloe-vera lip balm out of my handbag. 'Try this.'

Billy puckers rather than spreads his rutted lips. I smudge over the cracks as best I can, noting with a smile my father's inexperience at putting on lipstick.

Once out of his earshot, we discuss the parameters of his urine like the weather. It goes from drought to flood. In a few days, his wish for a drink of water is granted to excess. When his kidneys start functioning, my father has to drink five to seven liters of fluid a day. Finding something he can tolerate in such volume is a challenge: water soon becomes boring, cola is too fizzy, fruit juice too sweet, coffee too harsh and the tea-trolley too infrequent. Lemonade with lots of ice, left for a while for the bubbles to dissipate and the cubes to dissolve, is his preference, yet every fresh giant cardboard container from the food court is greeted with a disgruntled sigh. He knows that after he's managed to get this one down it will all too soon be replenished.

What goes in one end is reflected in what comes out the other. As his kidneys recover their capacity to filter the metabolic scrap from his blood, their product improves in color and intensity. What was a very dilute Miller Lite matures until his Stella Artois is every bit as rich as the next man's.

As we improve, the guilt of how little I have done for my salary over the past few weeks begins to bite. Worse, the annual review of progress at the Neuroscience Research Center where I work is fast approaching. In less than a month I have to find some grand strategic thoughts to include in my presentation entitled 'New directions in drug discovery'. I steel myself for the task, for there is much to talk about. There are endless new technologies on the horizon that promise to make drug discovery faster and more fruitful. Hundreds of new drug targets from the Human Genome Project loom large. Better ways of testing thousands of compounds a day for unique activities could be good too. But I need more than just technology. I need something different, novel, unexpected; something to make people sit up and think. If Billy can climb the purgatorial hill then the least I can do is to put a decent talk together.

I scan through *Nature, Science*, the *Proceedings of the National Academy of Sciences* and other prestigious scientific journals for inspiration. I think back to recent meetings with some of my colleagues. The last scientific conference I attended was in Glasgow, just over six months ago.

'Maybe all the big discoveries in neuroscience have been made, and anything left is just window dressing,' my friend and colleague Deborah Dewar had said between mouthfuls of macchiato in one of the many new coffee shops along the Byres Road. At one time, when we were Ph.D. students together, we might have debated for hours whether this was true. These days we argue less, yet something inside me desperately hopes she's wrong. I can't believe there's nothing more than details to come; that the world of biology has run out of amazing new discoveries. But how can we know? Predicting what remains to be discovered isn't easy and predicting the future isn't science.

My father knew Debbie. He introduced her to oysters! While I studied in California, Debbie lived in my house in London with my father in the supposed role of landlord—a commitment that amounted to little more than filing in poll-tax forms and footing the bill for repairs and refurbishment. On one occasion, my parents had met up with Debbie at a brasserie near Piccadilly Circus. At the end of the evening Billy slipped an envelope containing a wedge of crisp twenty-pound notes into her hand as reimbursement for new bedroom curtains. Always one to make the most of an opportunity, he recounted this as the night he had dinner with a gorgeous Scot, young enough to be his daughter, and it cost him only four hundred pounds plus expenses.

When we were students with permed hair, short ra-ra skirts and the occasional attendance at a Campaign for Nuclear Disarmament (CND) demonstration, neurotransmitter receptors were the cutting edge of science. We both worked on receptors involved in brain disorders. Debbie studied Parkinson's disease, while I worked on depression. We had no idea that within a decade we would be able to model on a computer exactly how they work, predict their shape, size and form—and all from the knowledge in their genetic sequence. We were to move into the era of molecular biology, to find that most genes are active in the brain. And from there neuroscientists have confidently been tackling the problem of how the brain evolves, developing into its final, finished structure, capable of generating memory and consciousness. Could there really be nothing more than details to come?

A few months after this conversation it turned out that we were wrong. Actually, we weren't exactly wrong, because scientists don't ever admit to being wrong, we just have to revise our original hypothesis a little. At university, we were taught that the price brain cells pay for being absolute experts in the art of communication is that they lose the ability to do anything else. Workaholics with no other skills, they can't even divide any more. Brain cells don't regenerate. Once they're gone they're gone—or so went the dogma that tried to persuade us students that drink and drugs destroy neurons that can never be recovered. But in 1998 those ideas were turned on their heads. The irreplaceable nature of neurons may turn out to be a myth. Early indications came from studying birds in the 1980s. Fernando Nottebohm and Steve Goldman from Rockefeller University in New York discovered that canaries can generate new neurons in brain areas important for song learning. Later, they found that songbirds generate new neurons every spring as they learn a new melody to attract their mates.

If birds can generate new brain cells, why not mammals? When scientists looked really hard, sure enough, they found them. By the late Nineties two groups had pretty good evidence for neuronal stem cells in shrews and monkeys. Even though we don't understand their role, finding them in humans was a significant landmark in neuroscience.

Fred Gage from the Salk Institute for Biological Studies in La Jolla, California, and his colleague the oncologist Peter Eriksson from Goteborg, Sweden, studied hippocampal tissue from five patients who had died of cancer. The cancer patients had previously been injected with a diagnostic chemical called BrdU, which is fluorescent, and as cells divide it becomes part of the daughter cells' DNA. Cancer cells divide quickly and so contain more of the diagnostic chemical. It was always believed that neurons don't divide, so they wouldn't be expected to contain any. But the transatlantic collaborators found BrdU-labeled neurons in the hippocampi of all five patients after they died. The BrdU could only have come from neurons formed during their cancer treatment—while they were adults not embryos (they ranged in age from fifty-seven to eighty-two)—illustrating that new brain cells are indeed formed in adulthood. This discovery offers hope that our brains might recover from the excesses of youth, and much, much more. In 1999, *Science* magazine hailed the discovery of such stem cells as the 'breakthrough of the year', describing them as 'a rare discovery that profoundly changes the practice or interpretation of science or its implications for society'. For scientists, who cautiously hedge their bets, speak in caveats and parentheses, that's big—no, that's *really* big—news.

Stem cells. Yes! That's what I'll talk about in my lecture. Stem-cell research has to be important for the future of drug discovery.

The idea that something I believed and understood to be irreplaceable, to be lost for ever, but that now has hope of regeneration seems the perfect subject. Stem cells are true embryonic cells and they develop along the same pathways as they would if they belonged to a fetus. The same steps, the same fate, just out of synchronization with normal programmed development. How like my father's progress to date: his organs are recovering in much the same order as they were originally made. He is going through the same sequence of development that we have all experienced, just out of time with the rest of his life.

At some point in the forty-seven doublings that turn one cell into a complete human being there are cues that signal some cells to form a brain, others to form the heart or lungs and so on. Very similar genes shape development across the animal kingdom—from insects to humans. One group of genes, called homeobox or Hox genes, guides the early development of the head and tail end of the embryo. This happens surprisingly early. By the time an embryo is just three weeks old the cells that will form the forebrain—all those interesting parts I've already described: the hippocampus, amygdala and the frontal cortex, to name but a few—have already received their instructions and are past the point at which they could become anything else.

Some less developed organisms, such as the urodele amphibians—newts and salamanders—never lose the ability to re-create their body parts. Exactly how they do it has yet to be fully worked out. Until stem cells became topical it was mostly assumed that their cells de-differentiate, that they reverse to an embryonic form, hence their ability to grow another limb should they accidentally leave one behind, say, in the mouth of a predator. Whether cells fully de-differentiate or whether a small number of stem cells remain in the adult limb has yet to be clarified, but during limb regeneration the Hox genes manage development just as they did the first time round.

Wouldn't it be great if my father was a salamander! Well, not a

complete salamander, obviously, but a few of those amphibian qualities would be much appreciated in his current circumstances. Imagine how much more quickly he would recover if stem cells could be
activated to replace damaged tissue. Don't worry about the kidney
damage, build a new one! Is it possible that the two sheets of embryonic tissue, the epithelium and the mesenchyme, which together
form the kidney might be persuaded to repeat the experience later in
life? And we wouldn't need to worry about his heart if stem cells
could be encouraged to differentiate into cardiac muscle and replace
the damaged tissue. *Homo sapiens* have very few sites of true regeneration (the liver, fortunately, is an exception). For almost every other
organ, the fate of cells—the organs they will become—is determined
in the first few weeks of embryonic development and there is no
turning back.

Reversing embryonic development might seem like science fiction but the first steps to make drugs that encourage mammalian
cells to return to an earlier, multitalented form have apparently already been taken. A group of chemists from Scripps Institute in California have discovered a small molecule they call reversine which
they claim can turn a muscle cell back to an earlier, more versatile
type of cell. Imagine the possibilities this conjures up for the medicine of the future when we might hope to regrow damaged tissue
with the versatility of a salamander. Until then my father must
manage with whatever support we can provide as his terminally dif-
ferentiated body does its best to recover conventionally.

With three weeks to go before my big presentation, I arrive at
the hospital, stopping at Burger King, which, on this occasion, is the
fast-food counter with the shortest line. I collect the requisite cup of
iced 7-Up as I have done every day for the last two weeks and arrive
at my father's bedside.

'Could you get me a water bottle, please, Ruthie.' I know by
now that Billy means one of the cardboard, boot-shaped potties as
opposed to a bottle of highland spring.

There isn't one on the cabinet next to his bed so I nip down the corridor and take a couple from the cupboard. I probably should ask a nurse, but I know where they are. Besides, ward C5 seems hugely understaffed. It's easier just to do it myself than to interrupt someone else's five minutes of attention. The regular nurses are busy handing out medicines at the other end of the ward, so best not interrupt. A buzzer goes off somewhere in one of the rooms on the left. No one seems too bothered. I still jump but most patients and visitors have become desensitized to the alarm.

A recently admitted hunched old man in paisley pajamas stands in the corridor, looking confused and bewildered.

I weave past the disorientated octogenarian and duck into Billy's room.

'Mr. Dixon. Harry. Where are you going?' I hear behind me.

I hand my father one of the cardboard boots and put the other on the shelf beside his bed before discreetly excusing myself with: 'I'll see if the dinner trolley has arrived.'

'My dad seems much brighter today, Sarah,' I say to the hassled chubby girl who is now mopping the floor.

'Yes, he has been out of bed. Doctors are very pleased!' she replies.

'Do you know what's for dinner?' I ask.

An apologetic shake of her black ponytail is the response.

'Nurse! Nurse!'

An urgent shout from an adjacent room gets her up off her knees. Humanity reverts to type in a hospital setting; whiners claim their justifiable right to whine while the brave-faced stoically carry on.

'Nurse! Nurse! The volume steadily increases like the ringing of an impatient mobile and Sarah disappears into the next room, cloth in hand, as I head for the stainless-steel dinner cart.

The dinner choices are liver and bacon or fish pie, followed by jelly, yogurt or banana and custard. I expect it's going to be fish pie and jelly but the assistant scans down her list and says, 'Bill McKernan—anything he likes!'

'Banana and custard? Really, are you sure?'

'That's what it says here.'

I rush back to tell the news.

'Here, Ruth—take this, would you?'

I carefully put the full container alongside the other two, lined up against the wall. A good 300 ml, I reckon. I would happily measure it and mark it on his chart myself but that would definitely cross the line from support to meddling. I'm sure I could do a better job, though. Yesterday's chart said four bottles and a total of 2.4 liters. I'm certain there was more, just from my own calculations. And it was all in the same biro and handwriting. That means that there was only one entry all day, despite there being three shifts. Even if he mostly slept through one, there should be at least two sets of handwriting, two different sets of numbers. They're guessing, not measuring. I don't think these figures are accurate. Maybe it doesn't matter any more. It doesn't measure up to my standards in experiments, though. Once you start collecting data it's important to see it through properly. How else could you compare my father's recovery with the norm?

I relay the menu choices to Billy.

'Banana? Did she really say so? Ruthie, you'd better check.'

I look out to the nurse's station. Sarah is back. She checks through my father's ever-burgeoning folder of notes. 'Yep, anything he wants now. Banana is OK.'

I collect his chosen meal and cut up the liver, spread a paper napkin across his lap and hand him a fork. Next to the bowl of thick, unnaturally yellow custard sits the small, spotty fruit. Proudly, I peel it and cut it up. Without touching the flesh with my fingers, I tip the chunks onto the custard and they float on the top, held in place by the generous surface tension of a thick skin.

Never have I seen a more appealing dish. Not that Billy is particularly fond of bananas, but this is a matter of personal achievement. Bananas are high in potassium; forbidden fruit for those with kidney damage and a trophy of recovery. Today is a big day; the day my father earns his banana!

He looks at it and smiles. With great pride he picks up his spoon, scoops out one small slice and slips it reverently between his lips as if he were sharing the 'body of Christ' at a Catholic mass. He pauses for the appropriate second or two, nods and then pushes the meal to one side. He is back in the community of the able-bodied; the big moment is over.

'How's work? How's the talk coming along?' he asks. This is more like my old dad. We used to talk science when I was at university. Mostly I would complain how much more biochemistry there was for me to learn. 'Yes,' he would agree, tongue in cheek. 'It was all phlogiston and Bunsen burners when I was a lad.' Now, with time on his hands and his brain in better running order, he shows the same curiosity and interest that he did twenty years ago.

'Well, I've made a start, but there's still a long way to go. I've done a lot of reading and I'm beginning to get a plan together.'

The fascinating thing about stem cells, I explain, is that they self-renew, retaining the elements of immortality throughout life. When they divide, two cells are formed, one a stem cell identical to the original, while the other is slightly modified, one step along the pathway to its ultimate fate. For the first few days of an embryo's life all cells are stem cells; the corpulent world is their oyster. Gradually, over the course of nine months or so, their choices become fewer and fewer. The hope is to harness that early potential, to grow stem cells under laboratory conditions and find ways to nurture them into particular cell types or maybe even into replacement organs for the health of mankind.

My father thinks about another piece of banana then changes his mind. He flicks his hand by way of instruction for me to take the tray away. 'I'll have a cup of tea, when they come round,' he says.

Isolated from his normal world, my father has lost his sense of daily rhythm, so my family and I have become the cues that restore it. My mother is 9 a.m., I am 5 p.m., and at 10 p.m. each evening Andrew arrives to make Dad comfortable and 'put him down for the night'. If ever my father had considered Andrew inferior to his other sons because he had fewer academic qualifications and a more rebellious nature, my brother's nightly visits would have made him rethink. Devotion beats a diploma hands down, every time, whatever its outer casing.

My husband brings in a newspaper when he visits. If Billy is too tired to concentrate on the television or radio Gerry provides a commentary on world events. Scotland's 2–1 defeat at the hands of Brazil in the first round of the World Cup may have been written up in the sports pages as a poor game, but, in the world according to Gerry Dawson, the old fellow's countrymen played on winged feet; a legendary performance of national pride. Other visitors bring grapes, herbal drinks, a *Telegraph* crossword book, golfing magazines; every possible source of stimulation to optimize my father's recovery. Keeping him happy outside his normal surroundings is not easy.

Growing and maintaining stem cells outside the human body is not easy either. Scientists have worked out by trial and error which nutrients are required to keep cells alive in a Petri dish, how the cells' cycle is maintained and which stimuli direct their fate. Several have worked out how to make stem cells into brain cells of most types. When James Thompson from the University of Wisconsin and John Gearhart of Johns Hopkins University independently isolated and grew fetal stem cells in a lab, they generated the raw material with which to treat unlimited disease. This all sounds like a great new future but, as Auden wrote, for every skyline we attain. . . .

There are still many hurdles to overcome. First, uncontrolled stem cells can cause cancer. In one rat experiment, where fetal stem cells were injected into the brain, a small proportion of the animals developed brain tumors. Even one tumor is too high a risk when considering using stem cells to treat human brain disorders.

Second, where do you get stem cells from? Aborted human fetuses? Early human embryos? Umbilical cord? The ethical and legal issues are still a matter of great debate. What about the few stem cells that last into adulthood? If we could recognize and separate them from other cells, would they provide an endless source of starting material? Or would we be better off working out how chemically to de-differentiate mature cells back into stem cells and start from there?

Furthermore, the legalities of who owns what and how the profits from any major treatment will be divided have yet to be resolved. The American corporation Geron funded the original research and holds many pivotal patents. Other companies have sprung up in their wake, mostly headed by academics, each with their own proprietary technology and niche opportunity. Sorting out the expected financial spoils is a quagmire from which only the lawyers are guaranteed to emerge in profit. I sketch out the options in my rapidly expanding PowerPoint file.

Progress in stem-cell research is not unlike my father's recovery. One big step forward but we don't know how many retrograde steps there will be. The early signs are good but nothing can be assumed. There are still infections—yeast infections, MRSA—complications of his time in Intensive Care. He may need reconstructive facial surgery because the flesh under his eye failed to heal, leaving his baggy lower lid red and drooping like a bloodhound's. The skin on his toes is still sloughing off in strips—long-term damage from the oxygen deprivation. And he can barely stand unaided. How much of his former self will return—enough for us to be able to look after him when he gets out? Will he be confined to a wheelchair? Was the dazed view of his bathroom the last he'll see of the first floor of 55 Audley Road? Maybe a few months recuperating in some kind of halfway house will see him right.

The very next day I arrive to find him sitting in the chair opposite his bed. He has gradually progressed from sitting up and lying down under his own steam to turning over unaided and then to getting himself into a chair.

'Watch this,' he says and he cruises the few steps to the washbasin and back, holding onto the sides of the bed for balance, in much the same way as my children used to roam around the furniture when they were one or two years old. Except that my father's steps are slower, more deliberate, and performed as though his legs might buckle under some huge weight on his back. Then he tries to let go. To walk out into open space when your body doesn't know how, or can't remember, is mostly a matter of confidence. For a small child it's a small step. For a fully grown adult the risk is considerably larger. Toddlers don't weigh enough to hurt themselves when they fall over and the force of gravity on a drop of two feet is not significant. It's all about being the right size, as the physiologist J. B. S. Haldane wrote. 'If you drop a mouse down a one-thousand-foot mine shaft it walks away. . . . A rat is killed, a man is broken and a horse splashes.' I worry that if the head of a frail sixty-five-year-old falls six feet and two inches to the floor it could be lethal. My Scottish friend Dr. Dewar works closely with the Neurosurgery Department surgeons at the Southern General in Glasgow. Their books are littered with fatalities caused by nothing more complicated than someone tripping over and cracking their head on the granite pavement after a Saturday evening in the pub.

Two more days of physiotherapy and a Zimmer frame provide the confidence that gets my father to the door of his room. He pauses for a moment, his face crumpled with effort as he takes in the bustling activity of the nurses' station. He stares past the rows of concertina curtains to the heavy wooden door beyond as if he were looking out on a breathtaking view of the ocean. He sighs, turns like a tanker and returns, shaking, out of breath, to the security of his bed.

'It's all taking so much longer than I thought. Everything is such an effort,' he says.

My father could never be described as a patient man. Whirlwind, dynamic, motivated, leader, yes. Patient? No.

Perhaps a change of scene would lift his spirits. With the matron's permission we plan to surprise him with a family picnic—well, as much of a picnic as you can have on a concrete patio behind the fast-food outlets.

Dad's friends Brian and Vanessa, proprietors of the Starr in Dunmow, prepare a *bona fide* picnic hamper that Andrew collects on the way to the hospital. 'We are just going to take you downstairs for a breath of fresh air,' we say, helping him into the wheelchair. Mum pushes while I scurry around, opening doors and repositioning the belt of his dressing gown to prevent it from catching in the wheels. Meanwhile, Andrew takes the homemade pork pies and wild-mushroom quiche out of the wicker hamper and lays them out on Brian's best starched linen. Ian opens a bottle of crisp Chablis and helps himself to a glass. The components of a perfectly romantic picnic on an uncommonly hot day in May are all present and correct, with the possible exception of a boater, a blazer and some cucumber sandwiches with the crusts cut off. Billy will love it. Progress is assured.

But when we steer the wheelchair into the lift my father is confronted by his image in the full-length mirror. He takes in the tiredness of his heavy frame, the chipped yellow front tooth, the drooping grey skin under his left eye and angry red cleavage below. He makes no comment and his shoulders and chin drop a little further. We wheel him onto the patio in blazing sunshine, amid the other circled wagons and the smell of stale Burger King. Seven grandchildren and six adults all want to fuss over him. He is a man trapped in the middle of a racetrack. Information overload! His hard-fought smile decays as he declines the wine and resignedly sips his watery lemonade.

I dip a giant asparagus spear in the fresh mayonnaise and hand it to him.

'Did you know that some people can't smell the metabolite of asparagus?' I say as he nibbles on the tip. Dr. Juvenal Urbino, the hero of *Love in the Time of Cholera*, could. Gabriel García Márquez wrote how much he appreciated the 'immediate pleasure of smelling a secret garden in his urine that had been purified by luke-warm asparagus'.

'I mentioned this once at a dinner party,' I continue, waving my own thick juicy spear around to good effect, and the guests—scientists, of course—started arguing about whether the problem was that people couldn't smell it or they couldn't produce it. So we came up with a simple experiment: if each person ate asparagus and then provided a urine sample for the others to smell we could work out which was true.'

'You didn't really do it, did you, Ruth?' said my mother.

'No, of course not. Anyway, it turns out that the experiment had been done before.'

Andrew tips his head back and laughs so hard I can see his tonsils.

'I looked it up in the scientific literature. The *British Medical Journal*—1980, I think. It's the ability to smell that varies. If you can smell the metabolite in your own urine you can smell it in anyone else's, should you so desire.'

'How is it you can always find something unpleasant to discuss while we're eating?' says my mother.

'At least it's not the rat dissection again,' snorts Ian.

My father tries his best to join in the well-worn topics of family entertainment but from a more peripheral position than usual. Beads of sweat appear on his forehead. His trembling hand wipes them off with the stiff white napkin. Conversation eddies away until he is no more than an observer.

'Ruthie, take me back up to my room, would you? It's too hot for me out here,' he says at the earliest polite moment, trying not to notice the disappointment in our faces.

The next day when I arrive with the traditional cup of 7-Up the room is empty and my mother is standing in the doorway. Smiling, she nods her head toward the far end of the ward, where I can see a familiar outline of navy-blue dressing gown and maroon leather slippers shuffling down the corridor. The Zimmer frame has been downgraded to a stick and we admire a few more effortful, shuffling steps before I hurry to escort him back to his room.

'That's brilliant, Dad. You've made so much progress this week.'

My mother claps her hands and holds them into her bosom. 'Well done, Billy!' Her whole face is animated in exaggerated approval of the type she normally uses for reinforcing good behavior in her grandchildren. The rear view of my father painstakingly edging toward a public toilet is the most exciting thing I've seen in weeks.

Stem cells exist in only one or two places in the brain. You won't be surprised when I tell you that one of them is the hippocampus. After all, it is the happening place in the brain. When new neurons are born they can integrate into existing circuitry. All their uses have yet to be worked out: they might help repair the brain if it is damaged, they might be required for learning or they might have something to do with memory. As with any new area of science, what we know about is what the most curious chose to ask. In one area of research, Elisabeth Gould from Princeton University and her colleague Tracy Shors virtually eliminated the birth of new neurons from stem cells in the rat hippocampus by treating them with a drug that kills dividing cells. The animals had been trained to associate a sound with air being puffed onto their eye, such that each time they heard the noise they would blink. Without newly formed neurons they couldn't remember the association; stem cells seem to be involved in retrieving the information that one thing is associated with another.

The second place for neuronal stem cells is the olfactory system. This may be more important in other mammals than it is for us primates. Rodents, for example, have a more highly developed sense of smell, and their odor-detection system has some similarities to hippocampal memory. Rats learn the location of their nests, the changing seasons and the proximity of predators using smell, whereas we learn about our environment more from sight and sound. But olfactory memories, like any other, are laid down for future reference. Who cannot fail to recognize the perfume worn by a past lover, the smell of a baby's skin or the stale smell of school changing rooms past?

There is huge variety in our sense of smell, sensitivity to asparagus breakdown products being just one example. Smell works in a similar way to sight and neuronal communication. Volatile odors waft into the nose and bind to receptors on olfactory cells. Inhaled chemicals can be detected at less than one part in a billion. Each cell recognizes just one chemical and the pattern of cell activation is translated into electrical activity in the brain. Whereas we recognize color as a combination of three elements (the retina has distinct receptors for red, blue and green light) and the brain actually has many more neurotransmitters than just glutamate and GABA (of which more later) and consequently has a few hundred different neurotransmitters, the olfactory system boasts a repertoire of up to a thousand receptors, each activated by a unique chemical.

Olfactory neurons belong to that rare category of brain cells that regenerate. A rat olfactory neuron lives for about sixty days before being replaced. If the nerve fiber connecting the olfactory center to the rest of the brain is severed, all five million receptor cells are lost. Fortunately, the store of stem cells is sufficient to replace the whole lot subsequently.

We may not have the same capacity to replace olfactory neurons as rats but we can still learn new smells. The stale odor of the ICU waiting room, the too clean smell of the unit itself and the line of fetid containers, not unlike the smell of the telephone box outside the Catford Ram on a Saturday night, will be forever in my memory. Most smells are complex mixtures. According to Professor Reed, the human nose can smell up to 10,000 molecules, so a perfume that contains 200 or so chemicals is no problem. I laughed out loud when I read that a fragrance called DNA had been launched. DNA, a massive, twisted thread-like molecule, is completely odorless. In the terrestrial world, chemicals need to vaporize to be detected and even the individual components of DNA don't do that.

My father's intestinal flora is subject to great fluctuations, which is a consequence of the multiple antibiotics he is taking, exacerbated by the atypically wide range of species of bacteria on offer in a hospital setting. We can broadly tell when one microbe has ousted the previous colony by a change in the air. But we are complete amateurs compared with my father's favorite nurse, Peter, who can identify a gastrointestinal infection from the merest whiff of a passing bedpan.

Peter is a stout, lively Irishman immediately recognizable by his thinning ginger hair, hastily shaven chin and a regulation tunic so tight that you wouldn't give much for his buttons holding if he sneezes.

'How long before they might let me go home?' asks my father.

'Sure, but you love it here! Where else would you get this kind of service?' replies Peter. Getting a serious comment out of Peter is like getting blood out of an atherosclerotic capillary.

'Come on, Peter, you know the routine. How much longer do you think I will have to be in here?'

'I'll be happy to look after you for as long as you want and I'll not be going home till Christmas,' he says, ducking and weaving like Michael Flaherty on speed.

Everyone has a secret skill. Other than chatting endlessly without providing any information, Peter's is to name any microbe from the odor it produces—a useful party trick, and it wouldn't surprise me if he were to make a few bob on the side by betting on the results of the lab reports. His nose is as well developed as that of a master perfumer and arguably of greater value. His descriptions are as spot-on as the writings of an expert wine critic. Sweetly smelling of mocha and toasted oak, that's a 1980s Bordeaux to international wine critic Jancis Robinson; sulphurous and sharp with a hint of vanilla—smells like *Campylobacter jejuni* to Peter.

The flexibility of olfactory neurons might find other uses. Who would have thought that transplanting a rat's nasal tissue around an injured spinal cord could help an animal's damaged spine recover? But this is just what Professor Geoff Raisman from the National Institute for Medical Research in London has done. Improvement occurs up to two centimeters along the spinal cord, enough for the rats to reach out and grab a food pellet. The implications are obvious. If this approach can work for rats, perhaps olfactory cells can be developed into a treatment for the likes of the late Christopher Reeve? Two centimeters wouldn't get many paraplegics out of their wheelchairs, but it might restore some bladder, bowel or motor function, and the events of the past few weeks allow me a greater appreciation of what a tremendous benefit that would be.

My mother brings in a sweet-smelling stephanotis from the conservatory; a couple of days ago it was a sprig of forsythia from the garden. The eclectic collection of flowers in the plain glass vase on the corner table has grown into a hopeful bouquet.

'I can smell the lilac from here,' says Billy.

He moves the unread newspaper from his table and pulls a few cards from the pile underneath.

'Have a look at these, Ruth. They'll make you laugh.'

In each one of them Ileana Lugo's perfect copperplate script is unmistakable. Oscar and Ileana have known my parents for decades. Traveling companions, business colleagues, surrogate parents to each other's children, Oscar and Billy have shared most categories of friendship that life has to offer. Inside I read: 'Our best wishes from Bill and Hilary Clinton', 'Get well soon, Fred Astaire and Ginger Rogers' and 'yaba-daba-doo love from Betty and Barney Rubble'.

He watches my face as I read them and the reflective smile is a momentary twinkle of his former vigor, like he had at the sixtieth-birthday celebration when he took us all to Captiva Island in Florida or when he performed the Christmas Eve ritual of putting on his striped apron and chef's hat, pouring himself a glass of Springbank and cavorting round the kitchen, concocting hors d'oeuvres from his Cordon Bleu recipe books.

Past experience bears no comparison with the recent low ebb, except for one solitary occasion. It was Christmas Day in 1997 and my father had just sat down to carve the turkey.

'So, Bill, what has been the highlight of your year?' Gerry asked, hoping to hear about the latest business deal or maybe how Billy had beaten one of his rivals to a lucrative contract in the Far East.

There was a surprisingly long pause. Then, with a seriousness that rarely darkened his demeanor, my father said: 'It was when Dr. Marcus told me that my leukemia was stable.'

Leukemia! Stable! First any of us had ever heard of it. For me, a black abyss opened up. How could he not have said anything? Even if he didn't want to talk about it, why didn't my mother tell me? How dare they keep it to themselves! What is so special about being ill that they couldn't share the news? There must be something I could do? Was that why he'd been so worried about the tax implications of handing the company on to the boys? Random questions darted around in the darkness. God, did he think he was going to die? God, was he going to die? No, not yet, surely. People live for years in remission, don't they? My friend Ray's mother managed two decades in remission from breast cancer. In two decades Billy would be in his mid eighties. Anything else could get him by then. But how aggressive is leukemia? Treatments are progressing, aren't they? Is it worse or better than breast cancer? Think, Ruth, think.

I realized that I didn't know the first thing about leukemia. Except that Rebecca, at Brownies, had it and she recovered fully, albeit with curly, instead of straight, hair. What would Billy look like with curly hair? Perhaps more like his brother, who has curly hair. I suppressed a smile. Inappropriate at a time like this, not funny. I must look up Chronic Lymphocytic Leukemia (CLL) as soon as I get back to work. First thing.

'But you are going to be all right, aren't you?' I said, in desperate need of reassurance.

'Oh yes, love. It's stable. I'm fine.' Don't say any more.

Stem cells might be big news for the brain, but for the blood they are as novel as flared trousers. Bone marrow is the renewable source of our blood. It is nothing short of a stem-cell factory producing red cells, white cells and platelets—all the necessary components of blood—at the body's command. Seventeen million red blood cells are made each second and stem cells can step up production tenfold in a crisis. Stem cells from the bone have a restricted number of options compared with embryonic or some fetal stem cells; they can only make blood cells.

As if to prove the transient state of scientific knowledge, no sooner have I written this than I find the most elegant piece of work illustrating that bone marrow can not 'only make blood'. It can make many other cell types—even brain cells. Two separate groups did similar experiments. They looked into the brains of women who had died some time after having received bone marrow transplants from male donors. If they could find Y-chromosomes in their brain cells they must have derived from the male bone marrow. Eureka, there they were! Not many—in fact, less than 1 percent of the cells, even in the hippocampus, the most dynamic brain region—had Y-chromosomes and they represented many cell types, including neurons.

I hadn't thought, until now, that leukemia could be viewed as a disorder of stem cells. Normally, blood cells are formed in a highly controlled and organized fashion. Supply is tightly coupled to demand. If an infection is detected by the immune system, whose sole purpose in life is to recognize anything foreign, seek it out and destroy it, then more white blood cells are made. Lymphocytes comprise the white cells responsible for detecting and destroying cancer cells (T-lymphocytes) and for producing antibodies to detect invading microbes or particles (B-lymphocytes). Uncontrolled B-cell production is the more common form of Chronic Lymphocytic Leukemia, the type my father has.

Many genetic flaws can cause CLL. Most are single genetic mutations that happen spontaneously. Some are visible at a genetic level because they change the structure of chromosomes. Others are changes also associated with many other cancers, such as defects in genes that repair DNA after chemical or radiation-induced damage. The cause is not the same in every person, but the consequences are fairly consistent. Just one aberrant cell proliferates to flood the blood with exactly the same cell-type producing one particular antibody.

Excess B-lymphocytes compromise the immune system. Mature, antibody-secreting cells don't die but accumulate in the bone marrow, lymph glands and spleen. This happens slowly. Gradually, over years, the mature, cancerous lymphocytes crowd out other elements of the immune system. It's hard to fight an invasion armed with only bent peashooters. Amazing, then, that my father recovered from septic shock: his compromised immune system would surely have stacked the odds firmly against him. Fewer T-lymphocytes increases the risk of serious infection or of malignant cells infiltrating other organs. Declining numbers of red cells, white cells and platelets reflect bone marrow failure and in the later stages this causes bruising and bleeding disorders. It is a slow process and, while there can be myriad problems later on, early in the disorder most people have few symptoms and are well managed.

My father never commented again about his CLL. The information was public but, until his present illness, not for discussion. I suppose I could have asked him about it, but then I could have asked him about all manner of things . . . but I didn't. I wouldn't want to ask a question if the answer might make his disapproval public. His daughter should be perfect. So conversations took place without words. My father was the master of non-verbal communication and I spent forty years reading the gradient of his eyebrows, hearing opinions hidden in the timbre of his breath or spoken through the rattling coins in his pockets. So when our conversations skirted close to his illness I knew in the subtlest change of his posture and tone of his voice to take a different path.

At work, I immerse myself in the science of stem cells, deciding what is relevant, what will make it into my talk on new directions in drug discovery. In the evening I find out what I can about my father's progress but it is a much tougher job; information is written in illegible notes to which I have no authorized access rather than in kilobytes of documents, easily searchable in *PubMed*. The only way to know about my father is to ask direct questions—and lots of them. I interrogate whoever is on duty about the results of every test. How high is his potassium, his alkaline phosphatase levels? What about the CT scan, the results from those swabs? How high is his plasma vancomycin level today? Will he need another infusion? Such relentless monitoring is conduct unbecoming for a relative, I know that, but what does it matter if they hate the sight of me so long as my father survives?

The dissatisfaction that accompanies my father's growing strength is palpable. The better you are, the more tedious hospital seems. Too sick to leave, too well to want to stay, he is in no man's land. If only we could get him moved to a convalescent home or even a private hospital, anywhere to see a tree through the window and have a different filling in his sandwiches.

This is a problem I can get my teeth into. Dr. Marcus, my father's consultant hematologist, will be able to help. The difficulty is that Robert Marcus is very busy and almost impossible to track down. The last time we saw him he was quizzing a group of medical students as to my father's diagnosis. Their consensus that the lesion on his cheek was herpes amused my father greatly. There is something satisfying in being a difficult case and he seemed almost smug to have outwitted them as not one trainee mentioned septicemia.

I sit outside Dr. Marcus's clinic until the last patient has been seen. I am going nowhere until I have spoken to him. I can do most things I put my mind to: I changed the master cylinder on my car when I couldn't afford to take it to a garage; I traveled around Japan without speaking the language. How hard can it be to get my father transferred to a convalescent home? If it's a matter of money, we can find it. If it is logistics—tracking down a vacancy—I can make phone calls, visits, organize transport. I'm not leaving until I can go back to my father with some good news.

Robert Marcus explains gently, but firmly, that no private health-care institution will want to take his patient. If they take him, they will have to take his MRSA too and no hospital wants to risk that kind of deadly infectivity. Who wants to share a room with the uni-cellular form of Mussolini? Instead my father is going to have to go through the proper channels. He needs to be able to manage on his own and be medically stable before he can be discharged.

I brace myself to tell Billy the bad news, but it is unnecessary: he has read the outcome on my face before I even open my mouth to speak. 'It's OK, Ruth. I can wait. It doesn't matter.'

As in every good story the hero must pass three tests to be freed. If my father can walk up and down stairs, make a cup of tea and maintain a normal temperature for twenty-four hours the staff will be delighted to see the back of him. Having shuffled along the corridor to the stairwell, he manages to get down the short flight of stairs with the aid of a stick and, providing that he takes a reasonable rest in the red plastic chair on the landing, he makes it back up again too. The physiotherapist signs him off on the first attempt. One down. He talks his way out of the second test, bluffing that he hasn't made a cup of tea in years so why would he start now. If he wants a drink my mother will make one. Under any other circumstances she would splutter at the cheek of it, but these are not normal times. The second box is ticked. The final hurdle is the hardest because it is outside his control.

To supplement his compromised immune system he is given additional antibodies. The antibodies my father received would have been extracted from a surprisingly large volume of blood—probably several gallons. That's marginally less than President Jimmy Carter donated in his entire lifetime but far more than most adults even think about, until we're on the receiving end. The units of immuno-globulins are slowly infused over two to three days, but at the start of each infusion Billy's temperature goes up. Are the raised temperature and night sweats an indication of infection or merely a reaction to the foreign blood products? Twenty-four hours after the last infusion three different nurses arrive, each armed with an ear thermometer. Good news is a privilege worth jostling over. The bespectacled registrar willingly signs Billy's discharge papers when the red display flashes up 99.3 and a guard of honor lines the path to the lift and beyond.

I'm not there to see it. Timing is everything, and my presentation on the future of drug discovery falls on the very same day. I call my mother during the afternoon break to check on my father's situation. He is home. He is free. We have a future. I return to the meeting, quietly jubilant. All the effort has paid off. This is our moment of glory. Nothing will hold us back now.

I deliver an under-rehearsed talk on a subject I knew little about a month earlier. I speak with the aura of an expert, answer questions with a confidence bordering on chutzpah.

'So what exactly are you proposing? How do you expect to use stem cells to discover new drugs?' asks the World-wide Executive Vice-President (EVP) of Research, a man who oversees the work of ten thousand scientists, and corporate folklore tells of the day he summarily fired a couple of them for inadequately answering his questions.

'It depends on the therapeutic area you're interested in. I doubt that I would pick the same experiments as you.'

The room collectively draws in its breath. My words didn't come out quite the way I intended. It sounded too much like a challenge and nobody dares to challenge the EVP. You can anticipate his thoughts, remind him of the important scientific concept you were discussing last week, or at a stretch honorably suggest a different interpretation of the data. But no one shows less than full deference to the EVP.

My shot-from-the-hip elicits a time-expanding hush. I feel myself falling from grace and watch a reasonable scientific career flash before my eyes.

The EVP's laughter cracks the atmosphere, followed a microsecond later by everyone else's in the room.

I carry on talking, working my way round to recovery. 'We could try transforming stem cells, growing them on silicon chips, and screen for drugs that affect only those cell types.'

'How could stem cells be used in the discovery of more traditional, small molecule medicines?' the EVP continues, seemingly more engaged.

'We could use them to produce human liver and cardiac cells. As we all know, the two most common liabilities in any new drugs we develop are liver toxicity and disrupting the electrical activity that drives the heart's normal rhythm. If we could know that any new drug we develop is completely safe with respect to these two properties it would be a big step forward. There would be savings in time and money and we may need to conduct fewer safety studies in animals.'

The EVP seems satisfied. Later on at the post-meeting cocktail party, the head of administration sidles up to me and repeats a couple of comments made by the great and the good: 'The best talk of the day,' said one, 'That woman has vision,' said another, and my colleague elaborates further. 'They think you're management material, a strategic thinker with insight and guts,' he says.

I blush, more in relief that I have done myself no great damage. They are so off the mark. Management material? It's simply hard work, a lot of reading and the best I can do under the circumstances. The insight is no more than aspiration. And guts? I'm not the one with guts. My father is the one who beat the grim reaper in a corn-cutting contest. He fought his way out of Intensive Care, out of ward C5, and now he is back at home, where we can look after him. Me? No. I'm none of these things; I'm just a daughter surfing on a wave of confidence borrowed from her father's survival.

6

Mood

Months ago I dreamed of a tulip garden,
Planted, waited, watched for their first appearance,
Saw them bud, saw greenness give way to colours,
Just as I'd planned them.

Every day I wonder how long they'll be here.
Sad and fearing sadness as I admire them,
Knowing I must lose them, I almost wish them
Gone by tomorrow.
From 'Tulips' by Wendy Cope

A WEEK LATER AND THERE IS NO ANSWER WHEN I RING THE FRONT doorbell of my parents' house, so I slip along the passageway between the house and the church. The back gate is open, the sun loungers are vacant and my father is lying prostrate in the middle of the lawn, the walking stick in his right hand pointing up to some imaginary star. His lips gradually widen into a familiar, unbalanced grin, taking up a disproportionate amount of the left side of his face and revealing a cushion of red upper gum. He ejects a hearty belly laugh through the new gap in his teeth formed by his chipped upper-left one.

'What exactly are you doing down there?'

He ineffectively clamps his teeth together and snickers like Scooby-doo, knowing he has overstretched himself again.

'Well, I was trying to shoo a cat away from the begonias,' he says, as if that were explanation enough.

I laugh too, but then laughter is infectious. We can't help but return it. Robert Provine, Professor of Psychology from the University of Maryland, has studied more than a thousand examples of laughter. His analysis says that our range of forceful air expulsion, from demure titter to energetic guffaw, has little to do with finding things funny. In fact, less than one-fifth of laughs are evoked by humor. Most have more to do with making personal coalitions. Laughter is a sophisticated form of bonding. Think of teenage girly giggling or the backslapping hilarity in the bar after the soccer match. We rarely laugh alone.

I get my father onto his knees—no mean feat for a woman more accustomed to scooping up a three-year-old—as he carries on smirking with the self-satisfaction of a boy who has narrowly escaped being caught smoking behind the bicycle sheds. His susceptibility for getting into scrapes has not declined since the day my granny caught him sliding down the potato patch in his best school breeches and slapped his backside with a leather strap.

'I was fine. Don't worry. I knew your mum would be back from the shops any minute.'

I link my arm through his and guide him back into the conservatory. 'You just sit there out of trouble and I'll go and make us a nice cup of tea.'

'Not too strong, Ruth,' he calls after me.

It's so good to hear him laugh, to see him happy. This is the first time I've seen him smile properly in weeks and it is his own natural smile, warm and spontaneous, not forced by a desire to please those around his bed. I smile too, thinking of him sitting in the sunshine rather than lying in a sweaty, hard-to-handle hospital bed. How much nicer it must be to see an iris-edged pond rather than a noticeboard of get-well-soon cards and to have the sound of bees on French lavender in your ear rather than the breathy scratching of a plastic thermometer. My father was happy enough with ward C5 when the alternative was Intensive Care; now home looks so much better than hospital. Yet he is still very unwell and a long way from where he started. It makes you wonder where happiness comes from.

I asked several eminent neuroscientists that question. Maybe I was imprecise in the way I phrased my inquiry, because the responses I got were very different. The neuroanatomist pinpointed the occipital lobe for visual humor, the frontal lobe for recognizing things as intellectually funny and the motor areas that are needed for the physical elements of laughing. The behavioral psychologist said, 'It's all to do with positive behavioral contrast,' which in everyday parlance means that we feel happier when things are better than they were before. That would certainly fit my father's circumstances. But my favorite answer as to the root of happiness—and, sadly, I can't remember who said it—was 'Scottish country dancing'!

Happiness is our own choice. Many of us believe that, including Groucho Marx, who said: 'Each morning when I open my eyes I say to myself: I, not events, have the power to make me happy or unhappy today. I can choose which it shall be. Yesterday is dead, tomorrow hasn't arrived yet. I have just one day, today, and I'm going to be happy in it.'

We all surely have some degree of control over our mood, but as a biochemist I would acknowledge that neurochemicals have a big role in the way we feel, too. My father's sense of well-being (and mine, for that matter) owes nothing to the two primary neurotransmitters, glutamate and GABA, we thought about in earlier chapters. They allow neurons to talk at lightning speed and they are the black and white of the brain, the on–off switches; the collection of pluses and minuses that form our cerebral circuit board. If our brains were made of only these two chemicals we would still be able to do most things: form memories, wiggle our toes, respond to sound and touch quite adequately. But even slugs, snails and Forrest Gump have more cerebral texture than that. It is another whole family of chemicals, known as monoamines, that color in the monochrome mind. The release of any one of these three tiny molecules modifies the electrical firing pattern of the brain's principal wiring system. The result is to generate mood and feeling in our minds, provide us with optimism, depression and even the more nebulous sensations of reward and drive. (A further group of molecules, the endorphins and enkephalins, are also considered to be important in generating euphoria and pain relief. Their relationship with the monoamines is still being worked out, as are the subtleties of exactly what each neurochemical does.)

'You seem in good spirits, Dad,' I call through from the kitchen.

'Making progress. I managed to walk down to the newsstand yesterday. Mind you, it took me fifteen minutes. I bumped into Peter Hoare. He said it took him half an hour to walk that far after his op, so I'm not doing too bad.'

Monoamines work more slowly than the fast-firing circuits. The neurons responsible for their production are unevenly distributed in the brain, originating from just a handful of cell clusters in the ancient brain stem, deep below the buckle of your baseball cap. The nerves spread forward and fan out like a river delta, reaching most parts, and they release their chemicals in an aerosol-like dispersion rather than in closely coupled one-to-one synapses. Their presence at synapses subtly changes neuronal communication. Monoamines can intensify information as if they were political agitators or spin doctors and they can diminish its impact, like social workers and agony aunts. In this respect, monoamines are modulators; they modify rather than perform brain function.

Exactly how many monoamines there are is still a matter of some debate. If we think of excitatory glutamate and inhibitory GABA as black and white, then serotonin, noradrenaline and dopamine would be the three primary colors red, yellow and blue. There are other minor contributors like octopamine, histamine and tyramine and related modulators such as acetylcholine and adenosine. And there will no doubt be more. The Human Genome Project has found whole families of receptors for monoamines that we didn't even know existed until a few years ago. Unraveling what the newcomers do will provide students with Ph.D. fodder for many years, but for now those three shades are sufficient to create most human moods from the bluest depression to the heartiest laugh.

'Don't the hanging baskets look fabulous?' I say, putting down two mugs of weak tea.

'Yes. Your mother did a great job while I was in the hospital.' Billy moves his mug away and lifts a couple of envelopes from one of the tidy piles in the middle of the table. 'Here's something you can do for me. Go into the study and get my checkbook, would you?'

As ever, I oblige.

My father fills out the check and the stub before tearing off the former and slipping it into the only unsealed envelope. 'Something for the Intensive Care team. They were so good to me. Maybe they can use it for drinks at their Christmas party.'

The envelope is already addressed to Dr. Park and he licks the flap several times before running his fingers over and over the join. He puts the letter back on top of the pile and neatly exchanges it with the mug of tea as though they're chess pieces.

I look around while he takes his first sip. The Sun is incredibly bright, streaming across the pool. I can't look directly at the water without being reminded of our first color TV set when the horizontal hold had gone. The reflected light is softer, though, flooding the conservatory in a restful glow. The hanging baskets are at their best, stuffed with flame-colored geraniums and swathes of blue lobelia that trail down the wall. It's warm. It's lovely. The sweet smell of daturas fills the air. Yes, 'Happiness makes up in height for what it lacks in length,' as Robert Frost once wrote.

If only Billy could always be this happy.

'It must feel great to be home, eh?'

'It is, of course it is,' he says. Then he sighs the little sigh that more often accompanies the last elusive crossword clue and other things that irritate him, like knives put back in the wrong drawer, letters containing grammatical mistakes and tomato pips stuck to the kitchen counter. 'I know how lucky I am, Ruthie. Really I do. But sometimes I'm just angry I was ever ill.'

There's a debt to be paid for survival. He has spent hours like this, writing cards to friends, thanking them for their goodwill messages. He has warmly welcomed any and every visitor, however tired he might be. Gratitude is the easy part but, somehow, it hasn't filled the need.

Perhaps it's hard to accept that it happened in the first place and all the harder for not having anyone or anything to blame. You can't blame a bacterium. You can't hold something that small and insentient responsible. Even en masse they can hardly be held accountable. In my mind, it was the interaction of microbes and genetics that did for Billy and I blame no one. Well, 90 percent of me blames no one, the more judgmental 10 percent can find a small space in the dock for whoever admitted him onto the ward and couldn't keep an adequate supply of antibiotics pumping through his veins.

Besides, it is not my father's nature to blame anyone. The events of the last two months are not like other incidents when he would chide himself for his own stupidity and move on. Not like the time he was running around the swimming pool, slipped and broke his shoulder. He wore a sling for a few weeks, his shoulder mended and he learned to walk more carefully on wet floors. He is one to learn from an experience and move on. Whatever happens in life is of his own making. He hired that idiot, he'll have to fire him. If he wants to buy that company, he'll have to raise the capital himself. Billy, like Groucho, has never been a man to live by someone else's script.

I am so happy that he is recovering I fail to realize how differently we see things now. We both know that the possibility of septicemia, or worse, will hang over him for the rest of his life. There are restrictions, of course, precautions he must take to minimize the chances of additional hospital admissions. No more trips to Phillipi Creek in Sarasota for oysters (a greater sacrifice for my father than for many); no more drinking water of unknown origins; and he must be particularly aware of early signs of infection. If I were to think more carefully I might realize that some elements of life have moved outside my father's control. And that's a new and unsettling circumstance for him.

'It's all down to chance, Dad. Let's just be thankful you're still here.' I am too grateful for all the other possibilities that didn't happen. He can see. His brain is as sharp as it always was. He's back at home with his major bodily functions intact. 'How's the physiotherapy going? Still lifting the baked-bean tins?' I ask hesitantly, aware that I have over-scrutinized his state of health since his illness and my intrusive attentiveness is a habit so strongly reinforced that I'm having trouble giving it up.

He moves the mug away again and checks through the letters. 'How are the children? You should really be looking after *them*, you know, not me.'

'They're fine, Dad. Gerry's getting their tea,' I reply and make no mention of the whining that preceded my departure this evening. 'Why are you going to Granddad's again? You went to see Granddad yesterday,' Francesca had moaned, folding her arms fiercely, like the fifty-year-old wife of an errant husband.

Perhaps Billy is tired. Perhaps he needs to rest. Perhaps I should go home and tuck the kids up in bed. I can't remember the last time I smelled Francesca's wet hair and did up the buttons of her mauve fleecy dressing gown or read *The Dinosaur's Egg* to my small weary son. I bet Adam knows *all* the words by now. Anyway, I can tell I'm being dismissed, so I swallow the last of the cold tea and pick up my bag.

'Can you post these on your way home?'

'Of course.'

'I won't get up to see you out, if you don't mind. I'll see you on Friday at the concert. You are coming, aren't you?'

'Wouldn't miss it!'

I take the envelopes and when I bend down to kiss him there is something warmer in his kiss, something more intense in the way he sends me on my way.

'Drive carefully, Ruthie.'

He's never said that before. I've been driving for twenty years and I've been pretty much accident-free. Well, OK, there were those two minor accidents, but I had barely passed my test and that clapped-out old Mini had dodgy brakes anyway.

There is no inkling of irritation or urgency left in his manner. Each comment, each action, is tempered by a gentleness that lives closer to the surface.

That kind of calm contentment is controlled by serotonin (or 5-hydroxytryptamine, to give it its full chemical name, 5-HT for short). It is arguably the most popular neurochemical of the last two decades. Serotonin is abundant in biology, found in bananas, nuts and scorpion venom. It was extracted from blood and first named serotonin (from *sero*—meaning blood—and *tone*, as in tension) by its discoverer, Irvine Page, who noted that very low levels can contract blood vessels. Most of the ten milligrams or so of serotonin in the body is found in the blood or the intestine, but for a neuroscientist its actions below the neck are mere digressions. First and foremost, serotonin is pivotal to the way we feel and is therefore the most popular neurochemical in our heads. For the Prozac generation, serotonin is the all-time favorite, the 'feel-good' chemical. Millions of patients have been treated and billions of dollars have been made from drugs that increase brain levels of serotonin.

Its role in controlling mood was discovered by chance. In the 1950s, clinicians noticed that Iproniazid, a drug commonly used for treating tuberculosis, made some patients euphoric. In 1956, the psychiatrist Nathan Kline tested whether this could be used to advantage and gave the medicine to some of his severely depressed psychiatric patients. They improved! It turns out that Iproniazid blocks monoamine oxidase, the enzyme that breaks down serotonin.

Consequently serotonin levels rise and, in more poetic terms, the spirit does too. (Actually, monoamine oxidase breaks down other monoamines including noradrenaline and both serotonin and noradrenaline probably contribute to its antidepressant activity.)

Artificially raising brain serotonin levels improves mood, and the opposite is also true: lowering brain serotonin can make you depressed. Professor Phil Cowen at the University of Oxford and his group showed this by feeding volunteers a drink rich in all essential amino acids except tryptophan, the dietary component from which serotonin is made. Normal volunteers become depressed within a few hours after drinking the mixture, a response that is magnified in people inclined to depression. Tryptophan occurs naturally in many foods such as chocolate, red wine and cheese. Decades of preferring Jarlsberg to jam tart might not have helped my father's cholesterol level but perhaps it did his brain some good.

There are more effective ways of increasing brain serotonin levels than eating a high tryptophan diet. Normally, serotonin pours out from its net of neurons like morning commuters streaming through the barriers at Liverpool Street station. The mood-enhancing molecule is later sucked back into neuronal stores to be reused, like commuters getting back on the evening train. Imagine what Liverpool Street station would look like if passengers could not get back through the barriers—the concourse would be awash with people and some would overflow into adjacent streets. Antidepressant drugs such as Prozac precisely block the transport mechanism back into the nerves and keep the surrounding levels of serotonin high. That's how Prozac and other so-called 'selective serotonin re-uptake inhibitors' or SSRIs work.

Current antidepressants do practically nothing to those who aren't depressed in the first place. But is there something stronger? Is there something that could keep my father as happy as he was when I arrived? Serotonin levels can be even more dramatically increased by actively forcing its release. This is how ecstasy (chemically known as MDMA or 3,4-methylenedioxymethamphetamine) works. I'm not advocating it. Not only would something be lost in blending the height and length of happiness, but the long-term effect of changing dramatically one's neurochemical balance has yet to be worked through. Emerging evidence says it isn't all good.

No one is going to stop my father from attending the annual classical music extravaganza known to locals as the Thaxted Festival. It may not be Carnegie Hall but, in the backwaters of rural Essex, the fourteenth-century church of St. John the Baptist, our Lady and St. Lawrence is top-notch. Acker Bilk, Cleo Laine, national symphonies and various choirs have graced both venues. The concerts surpass even morris-dancing week (the week of celebrating traditional medieval folk dancing) as the highlight of the social calendar and each year my father's company sponsors one performance, entitling him to twenty front-row seats.

We would like him to stay at home and recuperate but there is no chance. My mother tried to discourage him from attending the Kent captains' dinner, but that didn't work either. She ended up driving him eighty miles, to Maidstone, to demonstrate to his golfing peers how well he was recovering. The increasingly dedicated golfer had just been elected captain of the Kent Captains' Association when he was admitted to hospital. Sadly, the previous two incumbents had died *in post*. On the face of it this might seem rather an ill-fated position, but, given that serious golfers qualify for the honor of captain at about the same time as they get a senior citizen's bus pass, perhaps not quite so unexpected. Even so, my father wasn't about to make it a hat trick. He struggled through dinner, thanked those present for their kind wishes, cut the jokes down to just a couple, and, unusually, none were aimed at the English. Then he turned in early, exhausted.

He didn't miss that opportunity to show he's still here, nor will he miss this one, so it's no surprise that when we arrive to collect him he is ready and waiting with polished shoes, fresh aftershave and a newly ironed hanky in his jacket pocket.

Inside the church, we take our seats in the front row.

Marijke Curtis, the festival's organizer, taps the microphone. All eyes turn to the regal vision of elegance in cream crêpe and black patent. Her hair is smoothly swept into a velvet bow. Her skin is even smoother.

'Welcome to the Thaxted Festival,' she says in box-hedge English. She goes through a few organizational details—drinks available across the road, tapes on sale, tickets left for a few later performances—and then she adds: 'I am delighted that tonight's performance of Prokofiev's *Romeo and Juliet* has been sponsored by a local company and I thank their chairman, Bill McKernan, for his generosity once again. Those of you who have been regular attendees over the years will know that Molecular Products has been one of our most loyal supporters.'

Marijke puts her manicured hands together and initiates a small round of applause. She catches my father's eye. He nods back and beams.

There is a scrabble of chairs, some coughing at the back and then a hush.

The moment the music starts, my father's expression becomes trance-like. He is as still as he was in Intensive Care. Nothing seems to move, except the Venezuelan conductor's petite frame, which sways violently, but only from the waist up and in perfect harmony with the violin bows around her.

Attention is complete.

We have noradrenaline to thank for that. Like other monoamines, noradrenaline originates in a few nuclei in the brain stem. The most important of these, the locus coeruleus (pronounced sir-oolyus), sounds like it could have been a character straight out of a Shakespeare play. In fact, the name derives from the Latin for 'blue spot' because that is how it first appeared, anatomically. The firing pattern of the locus coeruleus—and, consequently, the pattern of noradrenaline released—controls attention. When primates are not paying attention to anything in particular the locus fires in a slow rhythmic manner, like a metal detector bleeping regularly when there is nothing noteworthy around. This is scanning mode. But when we pay careful attention to one particular thing the firing pattern changes to a fast, rhythmic pattern, not unlike the violins' dramatic hiatus.

Noradrenaline is the brain's equivalent of adrenaline. In fact, it's chemically only one carbon and two hydrogen atoms different and the two neurochemicals work in concert. When adrenaline tells the body to prepare itself for fight or flight, noradrenaline tells the brain to 'pay attention'. One of the most direct routes for noradrenergic nerves is from the locus coeruleus straight to the amygdala. Fast firing in the locus means more activity in the amygdala—and that's the circuit that makes us remember things better when we pay attention.

We think faster, or maybe better, when we pay attention too. That's because noradrenaline floods into the cortex, our 'gland secreting thoughts', as the Nobel Prize-winning geneticist Sydney Brenner once described it. The human cortex is huge, and it is the size of our cortex—rather than the size of our brains *per se*—that sets humans apart from other members of the animal kingdom. When it comes to the hindbrain, the part that controls movement, respiration and other involuntary functions, our organ is nothing special, in fact it's similarly proportioned to the size of our bodies as in most other animals. The same is true for the brain stem. But the cortex is humanity's pride and joy. The highly folded outer layer of the human brain would more than cover the average family's dining table if it were spread out in one flat piece. In other mammals (with the exception of dolphins and large primates) it would barely cover a five-pound note. All monoamines diffuse through the cortex, putting context and texture into our thinking. A huge cortex allows us to think deep thoughts, put them in order and play them back for others to appreciate.

In the church, Prokofiev's *Romeo and Juliet* is the perfect piece for the occasion: love, death and loyalty. We are enraptured. And there is something special about strings in a church. On old stone walls the reverberation never quite ends. When the music stops and the conductor is again motionless, something is left out there. It isn't as clear as sound; maybe dust unsettled by the harmonics or a quiet wind across the walls. Like the brass plaques commemorating expired parishioners, some element of the event lives on past its time.

'That was absolutely the best concert we've ever sponsored,' says my father as we walk across the road for drinks and canapés.

'Would you like to meet the conductor? She said she'd join us for a glass of wine.'

In his capacity as benefactor, my father tells her all about the international reputation of his company, the countries they export to, including Venezuela.

'South America is one of the fastest-growing markets for Soda Lime, you know,' he says wryly.

Unsurprisingly, she doesn't.

'It's used in anaesthetic machinery to absorb carbon dioxide— and in submarines.'

My father has been to Venezuela. The three of us exchange stories of countries visited. We compare favorites and Billy edges ahead in the global stamp-collecting, but it is a close-run thing. Professional small talk continues for a canapé or two until weariness forces him to rest.

'So—how did you come to move to London?' I continue, paying little attention to the reply because all the time I am looking over her shoulder, scanning my father's actions. It's rude, I know. Continually watching him is a bad habit and, even though I know it's not appropriate, I can't shake it off. He sits down next to an old friend, propping his stick by the side of the chair.

'Good to see you up and about again, Bill. Bit of a shock to hear you were poorly,' I can just hear the Fifties BBC voice say, over the top of his paisley cravat.

I nod intermittently in response to the conductor's comments, all the time straining to hear what my father is saying. He is telling his dapper contemporary that I was doing voluntary work in the hospital and found him sick in bed. I looked after him. I all but saved him. He must know that what he is saying makes no sense! Everything he knows about me, and multiple previous conversations, must tell him this isn't true. He isn't making it up to flatter me; he wouldn't do that. He is absolutely serious and, though the memory is as unlikely as cold fusion, he knows it to be true as surely as he knows his name. He smiles in my direction. I am no longer a scientist, I'm bloody Florence Nightingale. And my patient is not as well as I thought. It's a good job we're going back to see Dr. Marcus for a check-up soon.

Should I mention my father's confusion? I wonder, as we drive to the hospital. What help would a hematologist be? It's not as though there's anything he can prescribe for false memory syndrome. Perhaps it would be useful for Dr. Marcus to know, though? No, I decide. It was an isolated incident. He has been pretty *compos mentis* since he's been at home, and other than his memory of his time in hospital, his facts have been neatly filed, as usual.

The balding, round-faced consultant greets my father like an old friend. 'Dr. McKernan, lovely to see you,' he says, stressing the word lovely. He warmly shakes my father's hand with both of his own and looks him up and down as if to say 'What a fine specimen of a man you are'.

We review the details of the past few weeks: the date my father left hospital, whether he will have some cosmetic surgery to tighten the baggy skin under his eye, what he has done since then.

'You went to a concert—marvelous!'

Dr. Marcus checks Billy's weight and notes it down alongside his other scribblings and then emphatically puts down his pen. 'I must say, you have surprised us all.'

My father smiles modestly. I swell with pride.

'Now, let's have a quick feel of those lymph glands and see if things have settled down,' he says as he gently strokes the side of my father's neck. 'And then we'll just pop into the examination room next door so that I can check out a few other things.'

'I'll wait here, shall I?'

'Yes, if you don't mind. We won't be long.'

The notes are right there, in front of me. The temptation is almost too much. Everything there is to know about my father's state of health is just across the table. I could: there's time.

What if they came back, though? My father would die of shame if I got caught. Nothing wrong with reading what's on the top sheet, providing I don't actually touch anything. I crane my neck and make out one number: 153. Is that weight in pounds or systolic blood pressure? The rest looks like Arabic. I can decipher almost nothing.

Dr. Marcus strides back in, followed more slowly by my father, who is still readjusting the collar of his sports shirt. The doctor sits down and writes a few more lines before looking up.

'I could feel some small swellings around your neck, under your arms and in your groin,' he says. 'Had you been aware of them?'

My father nods.

What! He never said! The doctor goes on to explain that these are the telltale signs of lymphocytes accumulating in his glands. Septicemia has taken its toll on my father's immune system. His chronic lymphocytic leukemia is no longer stable. The time has come for medical intervention.

'I think we'll start you on a little gentle chemotherapy,' says Dr. Marcus, confidently handing my father a prescription for chlorambucil. 'Gentle' and 'chemotherapy'—there's a contradiction in terms. Gentle might be appropriate for the more recent and exquisitely precise drug Gleevec, a highly effective new treatment for a different type of cancer. If Billy had to have leukemia, why couldn't he have Chronic Myeloid Leukemia (CML), instead of Chronic Lymphocytic? The mutant protein that instructs bone marrow to make excessive abnormal white blood cells in CML is unique. It is produced by a single genetic mutation and exists only in people with the disease. Gleevec exactly targets this mutant protein, making it safer and more effective than any other cancer treatment.

For Billy, there is no such specific treatment because CLL has many different causes. The chlorambucil my father has been prescribed has no such specificity; it's about as selective as Herod's henchmen. This cruder drug works by irreversibly binding to DNA—any DNA in any dividing cell. Cells with malformed DNA are destined for destruction by the immune system. The most rapidly dividing cells in my father's body are the excessive B-lymphocytes his bone marrow is pumping out, so proportionally more of these will be destroyed. And that's as sophisticated as chlorambucil is.

Ironically, one of the promising compounds I was working on as a potential drug for anxiety turned out to have very similar properties. But for this purpose, far from being useful, irreversibly binding to DNA meant the end of the road for that compound. Pharmaceutical researchers know that the FDA, the Food and Drug Administration in the United States that vets all new drugs, would not allow such a potentially toxic compound to be unleashed on the public for anything less than cancer, so its use as an anxiolytic would not have been approved.

Much of the drive back from the hospital is spent in silence. Then, as we near home, 'You'll have to help me find a bungalow for your mum,' Billy says. 'She has been on about moving to a smaller place now that you kids have all left home.' It is true that my mother has never been wild about their big house. It is right on the street, the traffic can be noisy and revelers from the bar on the corner are an occasional nuisance late at night. She didn't sleep well when my father was away. She had learned to like the house because he liked it; the big conservatory, drinks around the pool, room to entertain. It suited his sense of status.

'It hasn't come to that yet, Dad. There's plenty of time. You should think about the things you want to do. What about Florida? Marcus says there is no reason not to travel if you want to. We just need to find a hematologist who can monitor your treatment if you're going to be away for a while. You're only going to need to take the chemotherapy for a few months.'

'Scotland. I am going to go to Scotland,' he said, pursing his lips with determination.

That's more like my father, more like the cruise-planning, deal-making, uncompromising man I know. His more natural state is high drive, high dopamine. It is his background color, he wears it like schoolboys wear navy blue and widows wear black. Scientists might talk about drive, reward and motivation. All these mean dopamine when translated into the language of the brain.

Dopamine was found to have a role in normal brain function by the Swedish pharmacologist Arvid Carlsson in the 1960s. Since then we have learned much more about its role in motivation. In rats, when nerves carrying dopamine from the brain stem up to other regions of the brain are destroyed, they no longer show interest in eating or drinking. This extends to other normally rewarding pursuits too. High levels of dopamine, on the other hand, encourage the animals to keep doing whatever they were doing before. Amphetamine and cocaine both increase dopamine levels in the rat brain, which explains why animals that have been trained to press a lever to get a shot of cocaine will carry on doing so pretty much until they die. That's how reinforcing dopamine can be in animals.

With this knowledge it seems unsurprising that cocaine and amphetamine are so addictive. High levels of dopamine they produce lead to subjective feelings of well-being and increased motivation in humans too. Whatever is happening when dopamine levels are high is perceived as having value. Dopamine also helps form particular types of memories. It 'stamps in' reward-related memories in much the same way as noradrenaline stamps in attentional memories.

There are other similarities with the noradrenaline system. Wolfram Schultz, the Cambridge neurophysiologist, showed that the dopamine neurons fire slowly until an animal sees, or even anticipates, something rewarding, and then they burst into action flooding the amygdala and cortex with dopamine. This rewarding flood taken to extremes might underlie the persistence of gamblers against the odds. More generally it underlies a determination to reach our own personal goals, be they buying a house in Florida, making a million or writing a book.

Getting the right balance is as important in the case of dopamine as it is for other neurochemicals. More so, because this monoamine controls physical as well as mental drive and motivation. Dopamine initiates action as well as thought, as was so beautifully described in Oliver Sacks's book *Awakenings*. A group of patients who spent decades in a frozen trance-like state because of pathologically low dopamine levels in their brains were re-awakened by treating them with an experimental new drug, L-dopa. L-dopa works by being converted in the brain to dopamine, restoring the neurochemical back to normal functional levels, allowing the patients to smile, enjoy a conversation again or play sports. Dopamine's role has been illustrated in Parkinson's disease, too. When dopamine-containing neurons slowly degenerate and die, sufferers struggle to initiate movement. Careful cognitive testing shows that they can also be less flexible in their thinking and tend toward depression, finding life and all its pleasures less rewarding. Both the physical and mental manifestations of Parkinson's disease are improved by replacing dopamine.

For weeks I have thought of little else but my father's recovery. But now I must acknowledge the truth. With his underlying condition, he might have survived septicemia but the temporary recovery can be no more than a reprieve. The leukemia is eventually going to win. Whatever extra time he has is a gift; a gift from the generous blood donors who provided the antibodies he received in hospital and from Alexander Fleming who gave the world antibiotics. Both treatments will help to keep further infections at bay. Those gifts are for him and also for me. He's mine to care for now, and I'm going to make sure whatever time he has left is the best. It's my turn to be with him. My mother's had him with her almost all her life. From the day she first arrived home from school with the spotty fourteen-year-old carrying her books.

'Can Bill McKernan come in and wait for the bus back to Blair?' was my grandmother's first introduction to her future son-in-law— and all the more impressive because he lived a hundred yards from the school, while they were an eight-mile bus ride away.

My brothers have had plenty of opportunity to get to know their father. Ian and Andrew have seen him most hours of most working days for more than a decade. That's probably more access to his thoughts and opinions than they wanted and maybe a few less than he felt they needed. Stuart has moved to America—it was his choice. Yes, my brothers have had their chance. Now it's my turn. I've been so busy living my own life that my father has only been part of it by proxy. He paid my university fees, looked after my house while I was in America. Our recent relationship has covered the practical not the personal. Since he came out of hospital I've almost rejoiced that he was ill. Of course, I didn't want him to experience the pain of those three months and nor did I want the rest of my family to live through the anguish. But it made me realize how little I know him and that there is still some precious time left to try to fill in the gaps, to know who he is. For a man who is never short of conversation he wears his heart tucked well inside the pocket of his blazer. Does he believe in God? Who did he vote for in the last election? I have no idea.

But what can I do? I can take on the sorts of jobs that my mother isn't particularly keen on, for a start. The skin continues to peel off my father's hands and feet in sheets, like terrible sunburn—probably a consequence of damage through oxygen deprivation. Underneath his skin is more skin; not the tough hide he's built up over years but a frailer pink layer, vulnerable and raw. His health requires that the old is smoothed into the new. I can do that; I can lubricate his chaffed epidermis with Sudocreme. I can also clip his toenails; he can't bend down far enough to reach his feet and my mother has always had a bit of a thing about toenails. If anything should happen to her—not that I seriously think he has even an outside chance of outliving his wife—I would give up my job and look after him permanently. It's a daughter's destiny. I know what he likes to eat, how much water to put in his Scotch. I can be that person.

Whatever I did for him in the hospital was only the beginning. Others can go back to work and think their job is done, but not me. I will do everything I can to make whatever life my father has left the best it can be. If that means taking longer vacations from work, so be it (and only my colleagues would know this represents a substantial sacrifice for me). If it means encouraging his physiotherapy, running errands, taking him to the doctors, I can do that. Gerry will manage to look after the kids. My husband is a very patient man and he adores spending time with the children. He is the one who would have had more. I reason that Francesca won't even remember that her mother wasn't there to put her to bed and Adam has always been a Daddy's boy anyway. There are plenty of families where the mother is away for days at a time. Their children aren't delinquents. I didn't turn into a reprobate or love my father less because he traveled a lot when I was young. Children are adaptable, I tell myself. Anyway, even if they are asleep when I get home, I'm always there with them for breakfast in the morning.

A few months are all I ask. It will be the twenty-fifth anniversary of Molecular Products soon. Billy and the boys have been planning a big party. I'll take him to Scotland. I'll make sure he gets to Florida. Please let him celebrate the company's success, let him see Scotland and Sarasota. If we can accomplish those three things, it will be enough.

I will make him happy. I will attend to his every need. I can, I will. Serotonin, noradrenaline and dopamine color my thinking too. These are the raw chemicals that paint our mood. A blend of these essential pigments produces the infinite variations in our frame of mind as easily as cadmium yellow, cobalt blue and red ochre make up the *Mona Lisa*'s smile. Without them, we would experience little emotion. They have shaped my thinking as well as my father's. Without that trio of tiny chemicals the moving sonata at the Thaxted concert, the sense of joy that Billy has survived, the vision of his smile beaming up from the grass would be bland facts, devoid of feeling, empty of love.

7

Connections

The wire which is coloured green-and-yellow must be connected to the terminal marked with the letter E or by the earth symbol. The wire which is coloured blue must be connected to the terminal marked with the letter N or coloured blue-black. The wire which is coloured brown must be connected to the terminal which is marked with the letter L or coloured red or brown.

NOTE: If severed from the mains the plug must be destroyed. A plug with bared flexible cords is hazardous if engaged in a live outlet.

Extract from the wiring instructions for a British domestic plug

SO, SCOTLAND IT IS. THAT MUCH IS IN MY POWER. I INVITE MY PARENTS TO join us on a cheap weekend break, flying to Edinburgh and driving on to Blairgowrie, the town at the foothill of the Cairngorms where they grew up. For the first twenty years of his life this farming region was the backdrop that made him Billy.

I know the facts. I know the basic details about his family, and when I say his family I must include my mother too, for they have known each other almost all their lives. They met at school and the bond formed at age fourteen has never been broken.

My father is the second youngest in a family of four—two boys and two girls—and his father, Tom, the son of a shepherd, worked for the Royal Air Force during the war and thereafter became the local postman. He died when I was a toddler no older than Adam and I've never noticed he wasn't there—until now, until I think about what Adam might, or might not, remember when he grows up.

Shortly after Billy left hospital he asked Andrew, in a rare moment of reminiscence, what he remembered about our grandfather. There was no hiding the angry disappointment my father displayed when my brother casually said he didn't remember his grandfather at all. Billy's reaction was hardly justified, since his youngest was but a few months old when Tom McKernan died. And when he asked me the same question I reassuringly replied, 'Oh, yes. I remember your dad,' and gave a rough physical description of him, but, in truth, the faint memories I have come from some old, undersized, black-and-white photos of Billy's graduation. They show a slightly built, neatly presented man with sticking-out ears, although didn't all men have jug ears in the Fifties? All I know of him is that he was a clever, quiet person, a lapsed Catholic, whose spare time was spent with his nose perpetually in a book.

My grandfather was absent for a fair bit of my father's young life. He spent the Second World War ferrying mechanical supplies or parts for tanks and jeeps around the country. For several years he was rarely home, except for the time he drove proudly into town in a tank and the boys in Blairgowrie crawled over it like ants on toffee, led of course by his sons, Tom junior and Billy. Actually, they weren't called Tom and Billy in their youth. My father was universally known as 'Wullie' and my uncle was 'Sandy'.

I have a much more distinct memory of my grandmother. Mary-Jane McKernan (née Mitchell) died when I was thirteen and her death was softened by the perception that she was old—which, at seventy-one, she wasn't—and the guilt-ridden but unspoken relief that at least it wasn't my other granny, on my mother's side. Mary-Jane was a tough disciplinarian who threatened us with the strap and put the fear of her Presbyterian God into me, especially when she caught me hobbling around the bedroom wearing my handicapped aunt's leg brace. She took us to church whenever we went home and was the keeper of the well-thumbed family Bible.

My father's hand-developed prints show her as a severe woman with thin lips, thin legs and thick tights. In both our memories she was always busy. In the kitchen she would have a pinafore over her faded lawn-cotton dress and pull the most fantastic food out of an empty pantry. Scrapings from the meat became stovies, a Scottish dish similar to corned beef hash cooked with potatoes, onions, and meat. Cold potatoes turned into light, fluffy scones, and her rhubarb and apple jelly was legendary. In the poverty-laden years after the war, my grandmother was a maestro of invisible patching and collar-turning to make my father's hand-me-down school uniform acceptable. And when life improved, the energy she'd spent cleaning other people's houses, working in the butcher's and taking in ironing was turned to handicrafts. She embroidered table linen in cutwork or crewel or counted threadwork, diligently following the instructions in her *Stitchcraft* magazines. My grandmother died when I was a teenager but I feel I knew her. And I can see some of her character reflected in my father: his drive, his 'excellence in all things', his eye for detail. He's even got her maiden name as his middle name—Mitchell—the same as mine. I can't surmise what he inherited from his father. An appreciation of books and education, perhaps, and maybe the idea that a family is one loyal, tight, united unit where periods of parental absence are accepted without question or threat.

How will my own father be remembered by my children when he's gone? I can't begin to think. At eight years old, Francesca will surely remember something of her grandfather, no matter how much more time he has. But will she remember him as a bad-tempered old man instead of the energetic live wire I know? Will I be disappointed if they don't remember him at all, if they don't love him as I do? Or maybe it's more important that he knew *them*, rather than the other way round? Yes, I've done the right thing in bringing the whole family to Scotland together. My children should get to know their grand-parents better before it's too late.

And there is still more for me to know. The bare facts I have are no longer enough. What was my father like as a boy and how did growing up in Scotland make him the man he is? How did he become such a patriarch; planning for years to ensure our education, our security and future with his handbuilt business and astute financial management? And how did he become such a patriot? He could be appointed chief cheerleader of the Scotland appreciation society complete with kilt, sporran, and skean dhu down the back of his sock. Such love of family and homeland must surely be hard-wired in his brain. Emotional memories of such strength can't be unlearned or overwritten. These associations are too strong, too fundamental to his thinking, and they form the framework from which all else develops; an undeniable consequence of the unique properties that brain cells possess.

Brain cells are not like any other cells in the body. If you have a pile of liver cells, you pretty much have a liver. Each one can do what a whole liver does, but on a much smaller scale. If you have a pile of neurons you don't have a brain nor can you hope to construct anything like one. Brains develop. They are complex systems. Not complex in the way that Francesca might complain that her homework is complex, but complex in that a brain can't be understood by scrutinizing its individual components. Imagine a newspaper picture. It's made up of different-colored pixels. If you collect them together you have piles of dots in primary colors, but not a picture, and it would be near impossible to rebuild the original. A brain is not a brain until neurons have organized themselves, evolving gradually to form something that is more than the sum of its parts. The miraculous thing is that our neurons build their own picture. The way our brains develop ensures that no two end up the same, because no two people are subjected to exactly the same experience. Even identical twins brought up under exactly the same circumstances don't necessarily fall off a garden swing at the same time.

The 400 grams of cerebral putty we are born with comes with all our neurons present, most made in the first four months of gestation but only the brain stem, the region responsible for respiration, heart rate and body temperature, is fully formed. If it wasn't we would be in serious trouble, unable to sustain the most essential physiological functions of breathing, swallowing and digesting. Some parts of the brain are no more than primed, ready to be modified. The neurons are all there, already planted, saplings in spring—roughly in the right place, with the rudiments of roots and branches ready and waiting on sun, rain and nutrients to bud, sprout leaves and flourish. In a baby's first year the brain, its most dynamic organ, more than doubles in weight and every day it uses up double the glucose we adults need. Why would a baby's brain need twice the energy when he can only do a fraction of what comes so easily to me or my father? The answer lies in development. A young brain needs fuel to organize, wire, re-wire and find the right connections.

Our parents provide most of the external influences in early development and those influences shape the development of the cortex, the corrugated outer cerebral layer that contains three-quarters of our brain cells and is the last part to form. The first twenty-four months sees a huge re-organization of existing neurons in different cortical regions and the process continues well into adulthood. During development, neurons compete for connections and having an excess ensures that all necessary connections are made. Individual neurons assemble into networks; networks assemble into systems and systems carry out functions like hearing or vision. Neurons that don't find an appropriate place in a network automatically die. The strategy of making excess starting material is more costly than 'supply on demand', but it guarantees the best brain.

The more varied and novel the environment the brain is exposed to, the more synapses are encouraged to form. Young rats reared in an enriched environment—with running wheels, tubes to climb through and new textures and smells—develop 20–25 percent more cortical synapses than their siblings housed in small single cages without extra stimuli. The early years are the most formative. St. Francis Xavier, the sixteenth-century Catholic missionary, knew that. He was ahead of his time when he said: 'Give me the children until they are seven and anyone may have them afterwards.'

Seven is such an important age. Well, it was for me; I was allowed to stay with one of my grandmothers, in Blairgowrie, for the summer holiday, all on my own. That was the rite of passage in our family. St. Francis Xavier and my parents thought alike; seven-year-olds were mature enough to be trusted. Six weeks in Scotland without my parents! Independence! Unlimited access to a piano and no brothers to annoy me!

Contemporary neuroscientists agree that the core of cerebral connectivity is firmed up in those first seven years. Yes, my father must already have had some influence on my children. This trip to Scotland will give him more opportunity to put something of himself in their little heads. It needn't be anything profound, for ordinary, everyday experiences shape neuronal development just as well and the more varied they are the better.

When we get off the plane my father rests, watching the empty luggage carousel circling endlessly, while Gerry and I negotiate the car rental. We have been organized enough to trade in some Virgin air miles for two weekend car rentals and now begins the careful negotiation to try to exchange the two minimalist vehicles for one spacious people carrier without having to hand over anything even distantly related to real money. Assorted bags are piled up on the trolley and topped with Adam, who, like any three-year-old, sees the new experience of riding through the arrival hall as king of the luggage castle as a holiday in itself.

In years past, flying up to Scotland was a luxury the McKernans could ill afford. When we went home it was usually in a borrowed, clapped-out old banger with my brothers and me sleeping in the back amidst empty chips bags, plastic bottles of cola and browning apple cores. My mother forced Quells travel-sickness pills down my throat only to see them again at the merest whiff of fumes from the first gas station we stopped at. But my parents assure me that what was not a good experience in my mind was the best of happy times in theirs.

I had forgotten just how beautiful the road is from Perth to Braemar. Just south of the Cargill smithy, an establishment long extinct for shoeing horses, the stonewalled fields drop away to expose a magnificent view across the River Tay. The sun is high and the light is clear as we drive up in the late morning, but I remember it as being even better in the evening when the grass on the hills is greener and the trees shine like acid-stripped copper, framing the distant iron bridge and the grey turrets of the Ballathie House Hotel.

My father stares out at the view. 'There must be a lovely view from that house on top of the hill,' he says, and the words echo back from my childhood.

'I think you say that every time we drive along here,' responds my mother. 'You know, we should have tried to buy that house.'

He nods and makes no further reply. Too late now.

At the single track, humpbacked bridge over the River Isla I blurt out, 'Slow down, Gerry,' automatically, remembering how my father used to speed over it deliberately when we were small, cruelly leaving our stomachs hanging somewhere out over the water. The sign of the River Isla and the approaching road will always evoke anticipation of an unpleasant yet thrilling feeling in me. That sort of joke was typical of my father. Billy wasn't the gentle, huggable type who would build me a doll's house or hold my skipping rope. He teased his children, made us laugh, led by example, set high standards and only indirectly praised our attainment. To mock us was to love us.

On we go, past a towering wall of shiny leaves that stretches for a third of a mile along the road to Meikleour. 'This is the famous Beech Hedge. It's over a hundred feet high,' says our understated tour guide for the children's benefit. Francesca seems more impressed than Adam, who is barely awake.

More quietly Billy adds: 'When I see this, I know I am home.'

We pass by Meikleour, the village where my mother grew up. It has a population of 140, a bar, a post office and a large herd of dairy cattle. The quiet backwater is forever part of my mother's identity. Now, whenever she achieves something significant, my father's seal of approval is to say: 'Not bad for a wee girl from Meikleour.'

It is only eight miles further to Blairgowrie (a proper town with a tea room, hotel, fish-and-chip shop and Chinese take-out), where we check into Gilmore House, our bed-and-breakfast accommodation, just around the corner from my surviving grandmother's new bungalow. There is time to visit her before we show the children the rest of the town.

My mother's mother pours out the tea and hands round the McVitie's caramel shortcake. When Granny was younger and more active it would have been homemade pastry cases filled with tinned cream and peach slices or scones fresh from the griddle, dripping with jam made from raspberries picked by her own hand and cooked in her own barrel-sized jelly pan.

'Help yersel's,' says the ninety-year-old and she makes no comment on my father's slimmer figure or the floppy fold of red skin drooping under his left eye. Here he can't disguise the most outward remnant of his illness with sunglasses as he has previously done, for there are few credible opportunities to wear them in the Highlands of Scotland, even in late summer. Granny neither questions his need to lie down nor asks about his illness. The reason is simple: she doesn't know. We've thought about telling her many times but always found a reason not to. 'Let's wait until we know if he'll survive.' 'Perhaps when he's out of Intensive Care—always better to be able to say something positive,' we rationalized. 'She'll only worry,' my mother said. And when Billy finally spoke to her, once he was home, he didn't bring it up either. Well, what sort of conversation would it be? 'What have you been doing, Bill?' 'Oh, not much, Chrissie. I was unconscious for five days, in Intensive Care for another week, and then I spent six more in the hospital . . . but I'm fine now.' No, that's too much information to give your ninety-year-old mother-in-law over the phone. Anyway, now that he's recovering, it seems pointless and the deception is complete.

Reasoning, planning and deception (however well meant) are mental functions that emanate from the frontal cortex. They contribute to the overarching quality we call personality. We know this by studying people with damage to that area. The most famous and often quoted is Phineas Gage, the railway foreman in New Hampshire who survived an industrial accident in which an unscheduled explosion sent a three-foot-long, one-inch-wide iron bar clean through his prefrontal cortex. He retained all his physical capabilities but the God-fearing, teetotal family man became profane, capricious, obstinate, and lost the ability to make decisions. Similarly, a thirty-two-year-old army captain in the Vietnam War was injured by a bullet that entered through his left temple and emerged through his right eye, taking out a large chunk of prefrontal cortex in the process. Like the unfortunate Mr. Gage, his personality became disrespectful, rude and impulsive, with the confines of reason and appropriate behavior melting away.

The brains of all animals develop in the same sequence but the development of humans' goes further. Whole new cortical areas have evolved in us so-called higher primates. The romantic-sounding Broca's and Wernicke's areas give us speech and language. They are named after Paul Broca and Karl Wernicke, the nineteenth-century neuroscientists who discovered that the left frontal cortex is crucial to language. The same regions don't exist, as such, in other animals.

During the first seven years of life the cortex is expanded into something the size and rough shape of a small cauliflower, benefiting our analytical skills and paralleling a child's mental development. By Francesca's age most youngsters can make a simple plan and execute it. By the age of ten they can formulate a hypothesis, test it and arrive at the best decision, inhibiting all other options; a ten-year-old can decide to keep a secret. By the time we are adults, we have developed the power not to be driven by our emotions, to act contrary to instinct. We can willfully inhibit what would be automatic, train ourselves to do things that most animals would be hard-wired to resist. We can rationalize ourselves into carrying out potentially harmful acts—you only have to watch someone poke their own fingers in their eyes while putting in contact lenses or avoid flinching from an anti-tetanus jab to appreciate that. As the psychologist Jean Piaget was the first to propose: 'Logical thinking is the highest form of biological adaptation'. Only we *Homo sapiens* are able to put reason above instinct or emotion.

At first blush, this all makes sense, until you think about bungee jumpers, skydivers and Bill Clinton's frisson with Monica Lewinsky. We know that we have the ability to reason what is best for us, yet sometimes, as Alexander Pope said in his *Essay on Man*: 'What reason weaves, by passion is undone'. How much sense does it make to jump out of a plane several thousand feet above the ground? Yet people do it just for the thrill. And whatever the President was thinking, sexual attraction is a very powerful force, stronger than reason.

Blairgowrie has always been a place of comfort for me; every trip is associated with security, freedom and family. Well, every trip save one. I was fourteen and very much a teenager. We had planned that I would stay with my grandmother for a few weeks during the summer holidays again. My iridescent turquoise platform shoes, Oxford bags and false eyelashes were already packed, although I had no idea where I was going to wear my trendy Carnaby Street gear. The day before I was due to leave, my father asked me to set the table for lunch while he and my brothers were watching soccer on television.

Why me? Why is it always me? I muttered to myself as I got out the salt and pepper. They never do anything—just sit there watching the TV all the time. *Singin' in the Rain* is over on the other side, I thought, as I laid out a selection of cold meats for sandwiches. But they won't let me watch that. Always the soccer, never the film! Those boys think I'm some kind of servant. A skivvy—that's what my mum calls it. Well, I'm not a bloody skivvy. Why shouldn't *I* watch the TV while *they* set the table? Well, I'm not going to do it this time. I'll show them. And I stood directly in front of the TV, deliberately blocking their view, and started throwing the knives and forks in my hand at the wall, one at a time, as if I were in a circus act. Clunk . . . clatter. One by one they fell to the floor. Clunk . . . clatter, clunk . . . clatter. Slap! My cheek smarted. My eyes filled with tears. It wasn't pain, just terminal humiliation. My father had never struck me . . . ever. I ran upstairs to my room in tears. I didn't speak to him that day, nor the next. I didn't say goodbye to him and I didn't speak to him the whole time I was in Scotland. And when I came back he made no mention of my bad behavior or his reaction, and he hasn't mentioned it since. Nor has he planted anything but kisses on my cheek.

It was the only moment of physical teenage rebellion I displayed toward him. The hormone-driven mood swings were otherwise re-pressed, or kept predominantly for my mother. I rarely pitted my will against my father's. We would argue, of course, but these were, at least in my mind, more matters of debate and less battles of will—the value of a higher education when I wanted to get a job rather than take up my deferred place at university, the rights and wrongs of the Gulf War, or the cause of Bovine Spongiform Encephalopathy (BSE) (before we knew about prions). Those conversations have dwindled as I moved through adulthood.

We park in the Wellmeadow but instead of going up toward the shops my father heads downhill to the river. This part of town has been gentrified since Billy was Wullie, even since I was a girl. The unkempt wasteland has become 'Riverside Walk and Gardens'; a grassy park with a trio of wine bars and a children's playground over-looking the river. The Ericht flowed higher and stronger through Blairgowrie before the Pitlochry Dam was built in the 1950s. But the water level has dropped and now exposes a river bed of rocks and carelessly strewn stones. The daffodils and carpet of wild garlic along the banks have been chased away by a dense canopy of trees. Their shade makes the brown water even darker and, though it looks like a torrent of English bitter, the smell is unmistakably peat; rich and warm and comforting.

For a man who has only recently relearned how to walk, my father heads upstream with more energy than a jumping salmon in spawning season. It is as though he has an appointment there and wishes not to keep someone or something waiting. I watch him charge on, myself delayed by the inseparable combination of a small child and a set of swings. I pause to push Adam a couple of times and watch his head tip back in pleasure before he hops down, falls over on the soft bark, gets up and toddles on after his granddad.

The contour of the river has changed. The banks were steeper and the water deeper when my father was a boy. 'I used to dive from there,' he says, pointing out the remains of a brick wall on the oppo-site bank to Francesca.

We follow Billy's feisty footsteps as he ventures farther to a set of wooden steps overlooking the narrowest part of the river.

'That's Cargill's Leap,' he declares. It is named after Donald Cargill, the minister who jumped the red sandstone gorge in 1638 to escape the troops hounding the Presbyterians out of Perthshire. 'People have died trying to jump it.' Failure would almost certainly mean falling into the rock-strewn water twenty feet below.

'Have you ever tried?' I ask.

My father raises his eyebrow as if to say, 'What! Are you mad?' and shakes his head. Even when he was young and adventurous he wasn't foolhardy. Someone he knew committed suicide from this spot. The river is to be respected.

Even so, boys of the Thirties and Forties misbehaved then as they do now. School holidays were spent swimming, fishing and generally making trouble along the Ericht's banks. Wullie and Sandy bombed salmon. They ignited a piece of photographic film, dropped it in a milk bottle with a little gas and threw it into the water: the fisherman's Molotov cocktail. I was fascinated and horrified by my favorite uncle's stories of how they picked the stunned fish out of the water and deliberately tore their mouths with a fishhook to make them seem honorably caught. This went on until the explosive flash temporarily blinded one of their friends and chemistry was returned to its rightful place in the school laboratory.

'There's not much else to see along here,' my father says, standing well back to let two hikers squeeze past us on the narrow path. He turns round, walks carefully down the wooden steps and sits on the nearest bench.

The water tumbles past, cosseting him in the sound and smell of his childhood. What has pulled him here? What is he looking for? Something is important, that much is clear. He is drawn to the river by a force greater than magnetism, stronger than gravity.

What is special about this river? What are you thinking? Who are you, really? I'd like to ask, but these are not the sort of conversations we've been used to. And more questions now would be an intrusion. Interrogation was justified when he was ill but even then I don't think I ever asked: 'What are you thinking?' Still, I'd like to know. How do I really get to know my father?

I open my mouth to speak before deciding what to say. 'Are you tired, Dad?' is what comes out. It isn't quite what I wanted.

'No, Ruth. I'm fine. It's just that Cargill's Leap is further than I remember. We'd better get on and do the shopping.' He does up his windbreaker and gets up, steadying himself on the arm of the bench. The moment is past.

There are set moments, windows of opportunity, in brain development—critical periods that affect the way the brain matures—

and information is only useful if it arrives at a particular time. Take the visual system, for example. Physiologists David Hubel and Torsten Weisel of Harvard University found that if you keep a kitten's eye closed during the first two to three months of life—a period that we now know to be critical for visual development—nerves connecting the eye to the visual cortex rewire dependent on experience. Without input they decay. Similarly, deaf children stop trying to speak or make any kind of vocal noise in later infancy because the development of language requires hearing and frequent sound input. During the critical period for hearing, sound stimulates hair cells in the ear that then activate auditory nerves. Neurons wired up to the cortical language centers will not endure without stimulation. By the age of six or seven the most influential period has passed, making it harder to teach a profoundly deaf child to speak. The brain is most plastic in the young. Many a neurosurgeon will tell you that brain injury can be managed much better by infants whose neurons have the greatest capacity to rewire.

For some systems the brain 'requires' consistent stimulation and allows a window of time for this to happen. This is the case for sight and hearing and touch, but what about our other cortical characteristics? What about emotional, rather than physical, development; the features that comprise our personality? Do the events of one's youth really provide the basis for the way we think for our entire life? Does a stable, routine childhood make for a well-balanced adult? Despite the war, the absence of his father and the relative poverty of the region, my father's youth was, by many standards, uncomplicated, traditional and routine.

The McKernans lived right across the street from Davie Park. There, Wullie and Sandy spent their evenings playing soccer or cricket, until dark. Then they returned home to read Enid Blytons and Agatha Christies by torchlight, under their bedcovers. On Saturdays they would go to a matinee at the Regal Cinema or Quinn's. The boys preferred Quinn's because jam jars or bottles were accepted in lieu of an entrance fee. On a perfect Saturday morning they would watch *The Corsican Brothers* all the way through, twice, and retrieve a couple of jars from just outside the back door on their way out, ensuring admission to the following week's showing of *Lassie Come Home*. On Sundays they always went to church. I am greedy for the details of my father's early years. Maybe they hold the key to what makes him special.

We join the Saturday morning line outside Irvine's, the butchers. The family-run sausage-and-pie business is an enduring landmark in the main street. Even in parochial, small-town Blairgowrie the tradition of buying groceries, meat, fish and vegetables in separate shops has mostly been replaced with one-stop shopping at the Co-op.

'Is this where you worked, Dad?' I ask.

'No, that was Petrie's, the grocer's shop, in Brown Street. It's gone now, though.'

Ah yes, I remember. I know a little about my father's first job delivering messages on an old black metal bike with a huge wicker basket on the front. For his fortieth birthday I put together a book of anecdotes and messages from friends, entitled 'Oor Wullie' after the cartoon character in the *Sunday Post*, and it was then that Bob Stewart, the assistant at Petrie's, told me about the time they worked together. Bob and Betty had one daughter but no sons, and whenever Wullie went missing he was to be found at their house. Bob took him fishing, fed him bacon rolls and lent him books. And my father, for his part, mocked Bob to the fullest extent that mutual respect allows. Bob once asked him to clean and polish a large set of scales out in the back room. Somewhat frustrated at the time it was taking, he told Wullie to get a move on or he wouldn't have them finished by six o'clock closing. After a while the stringy youth reappeared and silently carried on with his other chores. When Bob went to lock the back door he found the brass equipment reassembled, polished to perfection and covered with a white cloth on top of which a ticket read 'Unveiling tomorrow!'

We wait in line, eyeing up the white pudding, black pudding, venison sausages and mutton pies—things I haven't tasted in years, not since I was a small child and Granny visited us in London, bringing a selection of Irvine's best wrapped in a double layer of greaseproof paper tied up with string.

My father and I see the same landscape, eat the same food, but our experiences are far from similar. I've seen these red pavements and cold granite buildings before but they mean much less to me. Now, if we were passing the asphalt garden where I played as a toddler or the bus stop outside Harrold's sweet shop in Mottingham, London SE9—the locations I passed every day when I was a child—they would fire my emotional circuits. My father's white pudding supper from the chippie in Blair was my burger from the Wimpy Bar in Eltham High Street. The greatest luxury of his youth was a knickerbocker glory from Visocchi's, the Italian ice cream parlor in town. You needed the spoon's full length to get the cherry out of the bottom. For me, it was that very first double-thick banana milkshake from McDonald's on Catford High Street.

I'm collecting stamps: finding out small things about Billy and storing them away to look at again later. The album is filling up but I know I can never absorb all my father's experiences, so how can I ever truly know him?

The Canadian physiologist Donald Hebb first proposed how associations are learned, how memories are made. More than fifty years ago he suggested, using more specific and articulate language than I can paraphrase here, that 'neurons that fire together wire together', whereas those 'out of synch fail to link'. Synaptic connections are the material basis of our mental association. Our memories are 'everywhere and nowhere', as the American 'father of neuropsychology' Karl Lashley put it in the 1950s. Neurons—even from different parts of the cortex—assemble to create an image, a sensation or thought. There is not a single neuron that fires for ice cream or cherry or Visocchi's or spoon. Scientists once thought that each item or memory, the 'engram' as we call it, existed in a unique brain location, whereas we now believe that it is contained in a diffuse circuit of linked neurons. Location is still important. Neurons in the superior temporal sulcus (part of the cortex), for example, respond to faces. But there is no single neuron that defines my granny's features. More likely, she is represented by a unique pattern of neuronal activity in that region, with neurons firing for individual characteristics that together comprise her face.

'That's the school where your mother and I first met,' says Billy. 'Blairgowrie High!' He announces it fondly, with the warmth of a million neurons firing in synchrony. I imagine him as a star pupil—not from anything particular he has told me but from a collection of photographs of him in the local soccer and cricket teams, formally posed with arms crossed and slicked-back hair. Some even have a cup in the foreground. And I know he left Blairgowrie High with a big crop of 'highers' and 'lowers', the academic qualifications of his day—thanks, as he would say, to the benefit of a Scottish education. But then, my dad was smart in all dimensions. The first boy at Blair Atholl scout camp to find out that there was a deposit on the lemonade bottles, he collected them assiduously and with uncharacteristic tidiness. When the cricket team had an away match he played cards on the train with anyone who might have money to lose, including the conductor. The Scottish playwright J. M. Barrie could have had my father in mind when he said, 'There are few more impressive sights in the world than a Scotsman on the make.'

Billy worked wherever there was work to do: he earned money putting out the salmon nets, picking berries and pulling potatoes. Pulling potatoes was the best—you could always take a boilin' home, and his preference for Golden Wonders and Kerr's Pinks still endures.

At harvest time, Smedleys, the canning factory, was the hub of activity and, as teenagers and as college students, my parents manned the pea line. My father was the supervisor and my mother, already a serious girlfriend, loaded the crop, weighing the sweet legumes by the hundredweight and ensuring that they flowed evenly into the canning plant. The last truckload was usually delivered by 10 p.m., after which the equipment had to be cleaned for the following morning.

'Yes, and whenever there was a late delivery you always put it on my line. I've never forgotten that,' my mother says, half-smiling as they reminisce, driving past the road where the old factory once was.

'Well, you were grateful for the overtime.'

'But it was always my line, never one of the others,' she replies, reflecting the tongue-in-cheek bickering that so amused their friends back then. By the time they'd finished, swept up the spilt and substandard peas (which weren't thrown away but saved for the cheaper Woolworths cans) and cleaned the equipment a second time, it could be midnight. After the harvesting of the peas came the strawberries then the raspberries, starting at 6 a.m. each day until every field was stripped. Some of the students took Dexedrine to keep them going; others found places to snooze between deliveries. Billy's favorite was the twelve-foot-square vat that held the peas. He would crawl in through the opening at the bottom when it was empty and crawl out when he heard the sound of a tractor.

'Can you still get Smedleys canned fruit?' I ask, wanting to be part of the conversation and feeling perhaps a little envious. How wonderful it must be to have so much shared history, to look back on your whole life and know that the person by your side lived almost everything with you that you can remember.

'No, I don't think so, Ruth. The factory closed about twenty years ago when Smedleys became Smedley HP. It probably all gets frozen now.'

'Sometimes we picked the berries in the fields but it was back-breaking work,' says my mother. 'We'd get a penny a pound and keep going until our hands were blue. Dave the gaffer would shout out: "Make sure you pick your bottoms clean." How we'd laugh at that.'

We drive on past field upon field of rich red soil. There seem to be fewer than when my father was young but the raspberry canes are still so evenly parallel that you could write a man's life story between the lines.

Understanding how neurons that fire together actually wire together was an area of science that lay undiscovered until my father was in his forties, until technology was invented that could measure the electrical activity of connected neurons in the brain. Donald Hebb first proposed that persistent activity across a synapse leads to long-lasting changes in both nerves—that neurons that fire together wire together—but it was the British physiologist Timothy Bliss who proved it. In 1973, Bliss and his co-worker Terje Lomo put electrodes into just one of the millions of neurons contained in a small chunk of rat brain and recorded the current. When they electrically stimulated the cerebral region with rapid, high-frequency stimulation, they recorded long-term changes in the neuron's characteristic firing pattern; changes that lasted for hours, if not days. This phenomenon is known as long-term potentiation (or LTP). It is the mechanism that links neurons together and, since one neuron can be connected to hundreds of others, it's easy to see how huge assemblies can form, how the brain learns.

Think of driving to work, or a route you know well. Of all the roads you could go down you choose the same ones repeatedly until traveling that route becomes automatic. It can become so automatic that, on occasions, when you want to go somewhere else, you find yourself following the well-practised path instead. Or, when you look up, you realize you're further through the route than you expected. On a map the roads look no different, but in your mind the route has becomes one circuit, one united assembly of roads. So it is with the brain: through practice—through LTP—neurons link together to form a well-honed circuit and that assembly carries out an action or represents a memory, even a thought. One group of neurons can take part in many thoughts, just as one road can be part of many routes.

The elements my father holds most dear variously mesh together in the way that roads become a route and words assemble to tell a story. His love of Scotland and his strong family values don't spring from a single white pudding, from one hard-fought soccer match, a knickerbocker glory, a summer spent canning peas, or the sixpence hard won from playing cards with the conductor on the Blair to Kinross express. They originate from the wealth of experience generated in every part of his cortex, from a multitude of well-honed circuits, accrued slowly over his entire youth until the love of Scotland is so hardwired in his brain that my father is as likely to think differently about his homeland as I am to forget the way to work. Experience makes us who we are and so it's easy to appreciate why each of us is different—and why the brain is Woody Allen's second favorite organ.

When we stop for Sunday lunch at the Michelin-starred Peat Inn as planned, my father sits heavily in the only dining chair with arms. He seems tired from a long morning navigating us through both the local countryside and his early life. After the main course he dabs at the corners of his mouth and puts his napkin down on the table. 'Absolutely worth one Michelin star,' he says. 'Let me know what the desserts are like, would you? I think I'll just have a little rest in the van,' and he pushes himself up out of his chair.

'Well, your father never has been one for desserts,' my mother says as hollow justification once he's left the room. But still, his empty chair looks out of place and the fruit tart seems a little disappointing compared with the moist, thin slivers of pink duck we raved about earlier. We find room for it nevertheless and after the coffee I can't wait to go to the car park to find my father. Is he ill? No, he has just had a nap. He readjusts his seat upright, pops back into the restaurant to exercise his Gold card and we continue on our tour of the highlights of Billy's youth.

'I helped build this, you know,' Billy tells Gerry at the Pitlochry Dam.

'And that's where I went to night school,' he says at the sign to Dundee.

'And over there is the British Jute Traders' Association, where I got my first job and after a few months the boss asked to see my parents and told them my talents were wasted and I should go to university.'

'And this is the road I used to live in when I was a student,' he says in St. Andrews. And he points out the university buildings, too. Somewhere, in there, must be the biochemistry department where my father earned his first and the sports hall where he got his badminton blue.

Confirmation that these places still exist—and mostly unchanged—seems to bolster his confidence. Perhaps he needed reassurance that the landscape in his head still matches that outside. And as we walk along the beach, my father's steps in the sand are more like those of the finely honed golfer he used to be. The footprints point slightly inward again and he unconsciously rocks his weight from heel to toe. The flat-footed shuffle and stick have gone. As he looks over the waves, past the seaweed and detritus, the wind fills his clothes and blows them out to his once bear-like frame and he is closer to the father I know. He stands a little taller. The greyness of his complexion has been burned off by the wind and the way he thrusts his hands firmly in his pockets gives him a familiar air of conviction, defiance even. The image of a feeble man who couldn't cut his own toenails is gradually being replaced by the former, stronger picture.

Neurons that fire together are, thankfully, not necessarily wired together forever. Although our early years form the structure on which our lives are built, neuronal circuitry created by the strengthening of thousands of individual synapses can be modified. Synapses are malleable, changeable and plastic. Without synaptic plasticity we'd be unable to modify or build on circuits once they are formed. We'd find it very difficult to replace one image with a better one or change the route to work! And what about learning something new? If every circuit became fixed and hardwired, we'd be trapped by our early experiences, unable to rationalize why we are as we are and move on. Facts would be incontrovertible and we could never learn from our mistakes.

So how often do our cerebral circuits change? Just how plastic is the brain? These questions were answered very recently when time-lapse brain imaging (known as two-photon microscopy) was invented. You cannot imagine how exciting it is for a neuroscientist to be able to see neurons change and adapt as it is happening. Remember the first time you saw a film of a rose blooming using time-lapse photography? That's the way I felt when I first saw how dynamic neurons really are. At the Neuroscience meeting in 1998 a Californian company called Aurora showed a speeded-up video clip of neurons growing in a dish. They roamed around the dish, meeting other neurons. I could see them making connections. 'Sample and hold' is a phrase that's used to describe how neurons link up and it's actually what they do. I was stunned to see the cells synchronize. After several days in culture, when one was electrically active, all connected neurons were. A pulse of electrical activity spread around the dish like a Mexican wave at a soccer game.

Karel Svoboda used similar technology to look at neurons in a mouse brain rather than in a dish. He engineered neurons to fluoresce green in the portion of the brain that processes information from the animal's whiskers, monitoring activity through a tiny glass window in the mouse's skull. While the basic structure of the neurons—analogous to the tree trunk and major branches—remained constant, a small proportion of the synapses appear to change from day to day. Karel Svoboda calculated that 20 percent of synapses disappeared each day, to be replaced by new ones. When he trimmed the mouse's whiskers it experienced its environment differently and synapses split or branched more quickly. Further studies show that it takes only twenty minutes for a synapse to form. This work is still controversial but, if validated, it could mean that our neurons sample each other for about as long as we spend in a cocktail party conversation. If there is a mutual spark—synchronous electrical activity—the association holds and becomes meaningful. Are we permanently engaged in synaptic speeddating? In the time it takes to nibble a sausage roll and drink a glass of chilled white, the everchanging synapses in our heads could literally change our minds.

We consider heading back to the bed-and-breakfast but decide instead to stroll down onto the links. Billy wants to get a cap from the St. Andrews golf shop. The very proper nature of the place is evident in the white rails leading into the clubhouse and the occasional golfer still dressed in plus fours with thick wool socks tucked into spiked brogues. Their clubs are shiny and I can almost smell the thick leather of their golf bags. My father shakes his head slowly as someone misses a three-foot putt and for a while he is back on the course. There is a seedling smile, more than a hint of conviction. It was the right thing to bring him here. I can feel his resolve. He will play golf again. It's a given: that look of determination is back in his face. When he was just out of the hospital and in a particularly positive frame of mind he said that he was looking forward to learning how to play golf again. Adjustments would need to be made now that he was several pounds lighter but it would be something to aim for, a new challenge. Learning new things and forgetting old are both active processes. Even when cerebral connections haven't been used for years it takes energy to deconstruct them. Synapses have to be removed to allow new circuits to form, just as trees need regular pruning to bear fruit. It will take effort and training for the lighter, slimmer man to play golf as well as the behemoth used to.

Driving around Perthshire has been a tour of cerebral self-assembly. Revisiting the land of his youth shows that the memories are still intact, still happy and true. Trickling brooks and fern-clad banks pass by on the way back home as my father recounts his teenage fishing trips and the creel full of fish he ceremoniously dumped in the sink for his mother to clean. Somewhere in my imagination a romantic vision of a slim, handsome young man forms. Steering his bicycle with one hand, rods in the other, tea and sandwiches in the wicker basket at the front, my father is fast becoming a character straight out of *The Darling Buds of May*.

I came to Scotland to find out about my father, to understand what made him the man he is. How much is genetic and how much is environmental? Many books have been written on the subject of nature versus nurture—including some particularly good ones lately—and there is a mass of scientific literature testing twins separated at birth which explores this question.

It is becoming clear that nature and nurture are inseparable. The sequence of brain development is laid down in our genes. The part responsible for breathing and eating always develops before speech and hearing. Even the process by which the neurons wire together— the way they 'sample and hold'—is genetically determined and not influenced by environment. The final product, the individual organ that makes each of us who we are, is absolutely guided by the environment and we are genetically prepared to respond to it. Whatever happens in our lives shapes our neuronal architecture, the circuitry of our brains. Nature and nurture take equal parts in our formation. As the neurophysiologist Colin Blakemore once said at a lecture in Cambridge, 'Neuronal plasticity is a genetically programmed capacity to acquire information that isn't stored in our genes.' The way we respond to the environment makes us more than our genes.

Would Robin Hood make a Nelson Mandela if born in South Africa in the 1920s? In the right environment could a young Adolf Hitler grow into, say, Richard Branson? Experience is something you can't control. That is the utter stupidity of believing you can clone a person identical to yourself. Until the details of environment, particularly at critical periods, can be totally controlled, the only possible outcome of such a narcissistic quest for immortality would be serious disappointment. What would my father be like if he were born at a different time or in a different environment? He surely wouldn't be the man I know. A decade earlier and he could have been a war hero—or dead. A decade later and he could have been— who knows?—an IT entrepreneur, a .com millionaire.

If he had been brought up two or three decades later in Catford, southeast London, rather than Blairgowrie, Perthshire, would I recognize him? Would he be more like me? I realize now that I wanted to see Scotland as desperately as Billy did. I wanted to understand what exists here to make him so special. I thought this trip would help me understand him better. Why did he work so hard? Was it because of a poor upbringing? Why does he love traveling? Is it because he barely moved from a small town throughout his formative decades? How did this cold grey place shape him into something so special? It matters to me because what made him made me.

After spending a few days walking in his footsteps I know more details, that's true. I have a lot more stamps in my collection. But I realize, as I should have done before, that the way I feel about my father doesn't come from this kind of knowledge; from seeing the town where he grew up, picturing what he did as a boy, smelling the Tayside raspberries or even catching my own salmon. It comes from my own experience.

Like us, animals that are born relatively underdeveloped depend on their parents for survival. We are an altricial species like tree-nesting birds, rats or kangaroos and this drives our social behavior. Other species, such as ground-nesting fowl, chickens, ducks and quail, are precocial—meaning that they are born almost fully developed. After hatching, they follow the first thing they see. Usually it's their mother but it could be another animal or even the farmer's wellington boot. This behavior is known as imprinting, and ethologist Konrad Lorenz, in the 1930s, first suggested that imprinting was an instinct, a predetermined response to some elements in the parent, that develops during the 'critical period', the first twenty-four to forty-eight hours after birth. Imprinting may be less dramatically obvious for altricial animals but a parental bond develops nevertheless.

In humans, the bond develops more slowly and builds as we grow. The way my father held me up high in the air and dropped me down to kiss my head when I was a baby, the way he rubbed that Saturday morning stubble across my cheek, helped to build that bond. For two decades I wrapped myself daily in a green and orange silk kimono-style dressing gown that he brought back from Hong Kong. Even now, when it is threadbare, I still can't bear to throw it out. He set my standards. Whenever I've been tempted to rush through a piece of work I hear him say, 'If a job's worth doing, it's worth doing well.' And his words form the extremes of my vocabulary: guttersnipe, naysayer, zealot, Luddite. To use them is to think like him, to be him.

It doesn't matter where he grew up or whether his father was absent, his mother harsh or his younger sister handicapped. It doesn't matter that he lacked the luxuries of today's society; no pocket money, no *Simpsons*, not even tomato sauce until he was twelve. I can imagine his early life as though it were a black-and-white film, watch him grow up into a handsome young man with rolled-up sleeves and slicked-down hair freewheeling along the Scottish glens. But perhaps the answers to my questions are not to be found in Blairgowrie and his own past. My father is the way I see him because of the years of our shared experience. Others can't see him as I do. I love him because he is my father and I'd feel the same way if he were a wellington boot.

8

Genes

I am the family face;
Flesh perishes, I live on,
Projecting trait and trace
Through time and time anon
From 'Heredity' by Thomas Hardy

MOLECULAR PRODUCTS' SILVER ANNIVERSARY WAS TO BE THE BIG congratulatory bash that my father had dreamed of, promised himself, for years. Some might view a business as a purely financial asset, a tradable commodity, like stocks or bonds. The company he built was far more than a place of work. The product of his labor was more than an entrepreneurial gamble, more than a means to an end. Carole Francis was my father's personal assistant for seventeen years. Carole knew Billy as well as anyone and she summed up the energy that her long-time boss poured into the business when she referred to Molecular Products as my father's youngest child. It was as much a part of him as . . . well, as I am.

I was about fifteen when my mother called us all together and said there was something she needed to talk about. I'd heard that tone in her voice only once before, when my invalid aunt, who was staying with us, fell seriously ill. The ambulance men were preparing the stretcher to take her to hospital while my mother instructed us not to leave the sitting room until she returned. Her words were delivered with the somber cadence of a reporter at the site of a disaster.

Something imperceptible in her behavior triggered that memory and I saw again the indentation between her eyebrows, heard that oh-so-serious tone. Engaging all of us in one unified sweep of eye contact—something she probably learned at teacher-training college—she said: 'Your father is leaving his job and is going to start his own company.' Technically, it would have been more accurate to say that the job was leaving him.

Croda, the company that the ambitious young scientist joined over a decade earlier when it was much smaller and still called British Glues and Chemicals, was moving five hundred miles north, to Widnes. Billy—or Mac, as he was at work—had reached the position of Head of Gelatin Research, a job that included finding additional applications for the sticky bone extract used in photographic film and in glue on stamps. In a company where the senior managers were called 'sir', Mac lacked the conventional trappings of hierarchy and was the only head of department who would scrounge a lift to a social event without a pound note in his pocket, wearing a scruffy overcoat and a seven o'clock shadow. Nevertheless, his position had brought a reasonable level of comfort: a four-bedroom semi in the London suburbs; captain of Eltham Warren Golf Club; education at a minor public school for two of his sons; and a wife who was intimate with the layout of Marks and Spencer's, took pottery classes on Tuesday and had a wash-and-set each Friday. There was even enough money left for Friday night deliveries of Corona fizzy drinks in five flavors (limeade, cherryade, orangeade, lemonade and cream soda) and the occasional camping holiday in France.

'There won't be much spare money for treats,' my mother added. Actually, I wasn't aware that there had ever been any spare money for treats. It didn't matter much to me anyway. I already had a part-time job stacking shelves in a supermarket. Piling up the Carr's water biscuits just about managed to keep me in the teenage necessities of platform-soled shoes, cheap nail varnish and illegal Babycham (known as Genuine Champagne Perry) in The Royal bar at the top of Court Road. The magnitude of my father's decision completely passed me by. In self-absorbed adolescence I didn't expect it to affect me at all, so I continued in a state of moderate disinterest while the company that would become Molecular Products was born.

My father had a lot to lose. His first venture was a consultancy—McKernan Chemical Consultants—with a staff of one, providing services mostly to the United Nations, reporting on whether under-developed countries were ripe for setting up a gelatin-based industry. But, ultimately, my father wanted to create a company of substance, a lasting, solid manufacturing business with its own premises, prod-ucts and employees; a real entity that he could grow and nurture. Becoming self-employed was risky, but my father's drive for success was as intense an urge as any biological need to procreate.

The business was truly a product of my parents' combined ef-forts. In the early days, when my father worked from home with paperwork spread over the dining room table, my mother taught at Dorset Road Infants School five mornings a week and helped with the company books in the afternoons. When the business moved to its first premises, the company secretary continued educating her class of six-year-olds each morning and then hopped on the 108A from Eltham High Street to join the CEO in his office, an old coach garage, each afternoon. The business couldn't afford to reimburse her traveling expenses, far less pay her a salary. She did it for love. While their older offspring were getting a swift and necessary educa-tion in independence, my parents' youngest—the family business—was being nurtured and protected. Like the rest of us children, it was a product of their combined genes.

My father should know about genes. He was there around the time the biggest genetic discovery was made. As a biochemical re-searcher, Billy was a sideline spectator in the Cambridge glory days. We moved to Cambridge—actually a tiny village called Histon, a four-mile bike ride north of Cambridge—in 1959, when I was barely out of my baby carriage. It was just a few years before Frances Crick and James Watson won the Nobel Prize for 'cracking the code' of DNA and Max Perutz was similarly recognized for solving the structure of hemoglobin. My father was no Nobel Prize winner but by many standards he was a successful research scientist in the Department of Colloid Chemistry. After his first from St. Andrews, he gained a Ph.D. from Birmingham University, granted for exploring how to make an artificial blood substitute out of complex bacterial sugars called dextrans (well, it was a good idea at the time), and his best piece of research was published in the most prestigious scientific journal, *Nature*; an accomplishment that would be the pinnacle of any scientist's career. (It took me sixteen years to publish work of equal stature.)

When I asked him about life at Cambridge, his memories were not of vigorous academic debate in the Eagle but of evenings playing cards, working late in the lab and then compulsorily 'dining in', while at home his wife and three infants subsisted on eggs freshly laid by the local village hens. He grew vegetables and big-headed dahlias and didn't quite fit the mold of the ivory-tower intellectual. His commercial and entrepreneurial tendencies were a poor match for the Cambridge of the early Sixties. Had he been at Fitzwilliam today he would have been a perfect recruit for the blossoming biotechs of the Science Park, ripe fodder for the venture capitalist society, but at the time he was less interested in critical discussion of Crick and Watson's work than in the practical application of research and how he was going to pay the rent.

Like many of his contemporaries Billy underestimated the significance of solving the structure of DNA. As Max Perutz said when I met him the year before he died, few realized the true potential of what Crick and Watson had discovered. Who in their wildest dreams could have foreseen that the now famous sentence in their landmark paper of 1953—'It has not escaped our notice that the specific pairing we have postulated suggests a possible copying mechanism for the genetic material'—would lead to sequencing the entire human genome within half a century? Who could have foreseen the likes of antenatal testing, a murder conviction resting on the DNA in a single hair from the victim's shirt, or the promise of better antibiotics using the information contained in the streptococcus's genes? Who would have thought that, within the span of my father's working lifetime, humanity could advance from understanding Hardy's 'eternal thing in man that heeds no call to die' to the extent that we can begin to imagine how the tiniest variation in DNA might affect the contours of our personality?

The sheer extent of our DNA and the complex job that such an elegantly simple structure achieves is mind-boggling. Our genome contains a humongous amount of information; when decoded it adds up to three billion letters. If written in normal text rather than genetic code, even the smallest chromosome (that's number 21, which contains about 500 genes) would dwarf the complete works of Shakespeare.

Imagine if you had to provide a complete set of instructions to build a business from scratch; to make everything in the factory, from the boardroom table to the lab weighing balance, to the entire packaging plant. Just think of the reams of information you would need. Each brick, nut and bolt, even the recipe for mortar or the formula for magnolia emulsion, would have to be written down somewhere. We might divide the information into more manageable chunks—perhaps by using separate filing cabinets. Instructions would be written using an alphabet of twenty-six letters, formed into lines of text on a predictably huge mountain of A4 paper. It would be an unimaginable mammoth task in the same range of complexity as building a human body.

But building a human body is arguably easier because the information to make a whole person is contained in virtually every cell in our bodies. Each cell, enzyme, protein, even the recipe for mucus or the formula for tears, is described in our DNA. The information is divided into manageable chunks, stored on twenty-three pairs of

chromosomes. Instructions are written with an alphabet of four letters (A, C, T and G, each representing a chemical base). The sequence doesn't appear on a lined page but on a helical string of sugars
that form the supporting sides of the ladder-shaped DNA. Crick
and Watson first realized that, because the bases pair up across the
rungs of the ladder (A with T, C with G), the ladder can split right
down the middle, providing an elegant, asymmetrical template from
which two copies can be made. Each gene makes a protein (more or
less—there are caveats which we won't go into) and the segment of a
gene making a protein is 1,300 letters (i.e., bases) long, on average.
We do not fully understand what all our genes do yet but a complete
set of instructions for everything, from an eyelash to the antibody-
producing machinery of the lymphocyte, is written within them.

It takes roughly 25,000 genes to build and run a human being.*
We assume that's what it takes because that's how many genes the
Human Genome Project counted when the 'instruction manual' for
making a human was first mapped in outline form in 2001. Genes
take up less than 2 percent of our DNA. Interspersed between them
is redundant information, old crossings-out, and genetic doodles—
long trains of letters that mean nothing: they are quite literally
GAGA, or CACA. The text might once have been important but it
no longer matters. It's like the detritus that fills old filing cabinets:
yellowing memos and letters, expired business agreements and the
torn pages of the instruction manual from the original 1973
coffeemaker. Information gets discarded, duplicated and modified
on the road to evolution.

There are genes for structure, genes for function and even genes
that affect personality and behavior. Every person is, at least in part,
formed by their genes; structural and functional just as Billy's busi-

---

* The exact number of genes in the human genome is unknown. This is because we
are limited in our computational ability to predict them. Also, there are different defini-
tions of genes, for example, if we think of a gene as a recipe, is 'bread' one recipe or many?

ness is defined by its buildings and products. And then there are variations in genes that provide our personal characteristics—brown or blue eyes, red or brown hair, neurotic or placid temperament—perhaps in the way that the people a company employs provide the character and personality of the business.

The first building my father bought for the business lacked the grace of a body made from bone, keratin and muscle. It was a lower organism, a grubby, brick-walled shell in the slummiest part of Stratford, east London, and a place that the laws of hygiene and safety had passed over. Black metal barrels of chemicals, covered in a fine layer of industrial grime, hid enough rodent species to keep a taxonomist in gainful employment for a lifetime. The inappropriately named MP United Drug Company was actually a small chemical factory making just a few lab-based and industrial chemicals from a caustic raw material called soda lime. In my memory, if you could make it across the vermin-infested courtyard, you would find my father in the makeshift corner office, cigarette in one hand, phone in the other, amidst a jumble of metal filing cabinets, mounds of paperwork and overflowing ashtrays, intermittently talking to Bill Cheek, his foreman. Bill's wife, Joyce, a petite, Heather Locklear-type blonde with a beehive hairdo and cockney, salt-of-the-earth spirit, saw to the tea and typing. The Cheeks were my father's first and most enduringly loyal employees. To my father, MP United represented the early stage of a real business; solid bricks and mortar. He wasn't bothered by the squalor. To me, it was a dreary place I would not dream of showing off to my teenage friends.

Not for long, though. The business rapidly expanded and MP United's ugly period was satisfyingly short. Within three years the business needed more room and more staff. New premises were the company's metamorphosis and when MP United became Molecular Products Group PLC it moved to the southern edge of Thaxted, in Essex, a historic market town with a fine medieval church, cobbled lanes and a Guildhall.

It is still there. Some of the buildings date from 1520. The canteen was once a Tudor maltings. The block of offices was originally built in 1813 as a Baptist chapel. Throughout its existence, the main factory—the heart, lungs and guts of the place—has always employed the local townsfolk. Originally, they made old-fashioned mints, boiled sweets and barley sugar.

Now they make chemicals that absorb or produce gases, those most important to people: oxygen and carbon dioxide. The human body has evolved to function well at atmospheric levels of oxygen and carbon dioxide—about 21 percent oxygen and less than half a percent of carbon dioxide. When carbon dioxide rises to 2 percent it becomes uncomfortable, at 10 percent some people will panic and at 15 percent many will pass out. So removing carbon dioxide from the air in enclosed spaces like submarines and underground shelters is very important. Granules of soda lime efficiently absorb carbon dioxide from the air and Molecular Products makes several versions with colored indicators to show when they are nearing saturation. The technology definitely works. Billy tells a story of having sold a large consignment of soda lime to a Middle Eastern country for use in their underground nuclear shelters. The product's quality was tested by those in charge simply by shutting fifty people in the shelter for a weekend to see if they survived. Without 'sofnolime', rising carbon dioxide levels would cause occupants to panic, pass out and die of suffocation. Fortunately for all concerned, the experiment was a success and no one, other than Billy, suffered as much as a headache.

Similar chemicals absorb exhaled carbon dioxide from an underwater diver's re-breathing apparatus and are used in anaesthetic machinery. Molecular Products sells its wares to hospitals the world over. Do they use it in Addenbrooke's? I wonder. Was Billy the unknowing beneficiary of his own merchandise?

As the business grew, it expanded into making chemicals that produce oxygen. So when an airline steward announces: 'In the event of a loss of cabin pressure, oxygen masks will automatically drop down. Start the flow of oxygen by pulling the mask sharply toward you,' your sharp tug pulls the tab on a canister of chemicals and initiates a chemical reaction. Above your head could be a 'chlorate candle' similar to those produced by Molecular Products to provide that life-sustaining gas.

The business has become much more efficient: Sofnolime is made in one smooth production line, rather than as a batch process; a new automated machine packs it into handy-sized canisters, and pallets are stored in the warehouse ready for export. The parallels exist in human biology where, on a sub-cellular scale, a cluster of neuronal enzymes attach and remove atoms in a production line of five sequential steps to produce dopamine from its raw material, tyrosine. Thereafter it is packed into handy-sized vesicles ready for export out of the nerve.

As the business flourished, the staff expanded. As students my brothers Stuart and Ian regularly bolstered the workforce during their holidays. I myself worked at Molecular Products only once, for three long weeks, packing soda lime into five-kilo canisters. Six hours a day pouring, weighing, twisting, sticking and lidding. Frankly, that was more than enough to convince me that Molecular Products wasn't a long-term career option. Working in a chemical factory seemed a far cry from my university aspirations of single-handedly discovering new medicines, curing cancer and effecting worldwide nuclear disarmament. Nevertheless, it earned me enough money to join my friends in Europe for the rest of the summer vacation.

My father wanted and encouraged his family to join the business. Andrew joined the production team straight from school and Ian followed a few years later. My parents understood that neither Stuart, who was by then a physicist in the United States, nor I signed up because the company's remit fell outside our scientific interests. But underneath the accepted status quo I often wondered whether the timing just wasn't right for Stuart. Being the eldest, perhaps he was looking for a job before Molecular Products could afford to expand. For myself, in moments of sibling rivalry I claimed that 'I didn't need to go to my father for a job'. I meant it. I had inherited my father's independence. He built his own career and I would build mine. And there was something more holding me back, something so deep within me that I couldn't see it at the time, something about judgment and approval. Would I still be Billy's lovely daughter if we sat in adjacent offices every day? Would my father treat me differently as an employee? What if I didn't meet his expectations? Worse, what if he didn't meet mine? It's easy to appreciate the stories of his tough negotiating tactics and occasional loss of control in a board meeting but would I find it funny if I were the cause? And Billy seemed to know a lot about the arms industry—navy supplies, nuclear shelters, the workings of a submarine. As a CND sympathizer, perhaps there were things I wouldn't want to know? Just how deep into the whole industry could he be? Deep enough no longer to see his halo?

Anyway, with a degree in industrial chemistry, Ian was unquestionably going to be the next Managing Director (MD). Employees and family alike expected it and Ian's commitment to Molecular Products pleased Billy, ensuring his genes at the helm for future generations.

In a close-knit company, the employees were chosen as carefully as if they were marrying into the family. John Addison, a bearded, stocky bulldog of a man from Merseyside, has been the production manager for well over a decade now. He joined at the age of forty-five, despite his resolve never to work for a small business again. He once confessed to me, at one of my parents' many barbecues, that he was lured by my father's sincerity at the interview.

After only two weeks in the job, he pitched up in the boss's office with big demands. 'My staff needs a 20 percent pay increase across the board and I want their holiday allowance to go up from two to three weeks,' he said, talking as straight as a Roman road. He justified it by saying people would fiddle the money out of the company anyway. 'If you don't treat people right you can't expect their support. And if I haven't done the right thing—in a year you can fire me.'

Billy didn't, of course. He appreciated John's willingness to go out on a limb for what he believed to be right. He put up the wages, increased the holiday allowance, and on the rare occasions when he wanted a second opinion, John would always deliver it in the shortest possible distance between two points. His directness and fairness became part of the company's strength. But then a company's personality comes from its people.

Human personality, however, comes from our genes. It is important to bear in mind that there is no single gene for any one behavior. Nothing curls a scientist's metatarsals more than the inaccurate shorthand you read in newspapers about the 'gene for endurance sports', 'the gene for Alzheimer's disease' and 'the gene for aggression'. This implies that one person has a gene but another doesn't and that there could be a single gene for every attribute one cares to conjure up. It reminds me of the lady I overheard in the doctor's waiting room when I accompanied Billy for a checkup. 'I've got blood pressure,' she said self-pityingly. We all have genes, just as we all have blood pressure; it's the subtle variations in those genes that contribute to individual behavior in the same way that diet, weight, exercise (and genetics) contribute to blood pressure.

Jonathan Flint, Professor of Genetics at the University of Oxford, has been hunting for the genes that contribute to personality. He has focused on neuroticism (intensely anxious people) because this is a consistent and easily measurable element of personality. He sent questionnaires to 250,000 people, looking for pairs of siblings who were at extremes of the anxiety scale—both neurotic, both calm, or, best of all, one of each. Four hundred replies matched his requirements and those volunteers sent cheek swabs to Jonathan's research lab, where their DNA was analyzed. Some pretty heroic mathematics showed that many different genes contribute to neuroticism, with each contributing a very small amount—there is no single gene to make a man neurotic.

One likely contributor, though, is the so-called COMT gene (the code for an enzyme catechol-O-methyl transferase) which is responsible for inactivating the neurochemicals dopamine and noradrenaline. It comes in two forms that differ only in a single letter in their genetic code, making one form more stable and long-lasting than the other. David Goldman, Chief of the Laboratory of Neurogenetics at the National Institute for Alcohol Abuse and Alcoholism in Bethesda, Maryland, found that the more stable enzyme is consequently more abundant and so more efficient at breaking down noradrenaline and dopamine. People with two copies of the more stable form (one inherited from each parent) have lower levels of the mood-modifying neurochemicals. People with two copies of the *less* stable form have higher noradrenaline and dopamine levels and, in general terms, Professor Goldman found them to be more anxiety-prone and sharper thinkers, all other genetics being equal. I asked him during a break at a meeting in Bristol how he could tell which form of COMT we have. 'Coffee,' he said, raising his polystyrene cup, 'is a good indicator. Some people do well on coffee. It perks them up, helps them think. Others find that caffeine pushes them over the edge. They get irritable, nervous, shaky and can have trouble sleeping.' That would be me, I thought—but not my dad. I've seen him guzzle his way through a whole pot in a morning.

Simple genetic 'mistakes' underlie the infinite variety of human life. When the DNA-replicating machinery selects a C instead of a G, the consequences may be trivial or profound. The human genome abounds with single-letter substitutions, about one every 1,300 letters. These tiny variants, such as the difference in the COMT gene, are known as polymorphisms. Many are completely inconsequential, especially if they occur in DNA that isn't part of a gene. Some are detrimental; sickle cell anemia and other hereditary blood disorders are caused by a single base change in the relevant gene. Others, such as the polymorphism that sometimes occurs in the CCR5 gene, can be beneficial. The AIDS virus uses the CCR5 protein as a gateway into lymphocytes, where it replicates and destroys the immune system. People with a defective CCR5 gene develop AIDS much more slowly and rare individuals whose polymorphism means they make no CCR5 protein do not get AIDS at all.'

The portfolio of genes we start life with is our basic tool set. It influences our future from the seemingly trivial such as how we respond to coffee to the more serious such as how we deal with AIDS. The argument over nature versus nurture has been reduced right down to the demonstration that a single gene observably affects the way we behave or react to the environment.

The corporate body, Molecular Products, was also shaped by the material it started with—its founder, its building, its products and its people—and as it grew it was, like Billy, like any other person, shaped by its environment and experience. Sometimes the environment was good. The Gulf War, for example, generated a greater market for some of its products and the growing sales force fought hard for lucrative contracts to the American navy. And sometimes the environment was not so good: poor exchange rates hurt, as did high interest rates, and the company became well practiced in the art of reining in expenditure.

Building a business means taking thousands of decisions, some small—'Should we buy a photocopier before a filing cabinet?'—and some large—'Would a ten-year loan for a new production plant enable us to grow the business faster?' My father, like every other self-employed, aspiring magnate, actively took most of these decisions himself, particularly in the early days. In that respect, his baby Molecular Products wasn't like his other offspring; feed, clothe, nurture and the rest is automatic. The decision whether to build a skeleton before a fingernail, when to grow a heart or wire a brain, is genetically pre-programmed and requires no active intervention at all. Genes switch on and off in a defined order. And that order is largely the same from horseflies to *Homo sapiens.*

No—for Molecular Products every decision, the consequence of every small step, every order, every expense, had to be consciously thought through and planned. My father was more actively engaged in building Molecular Products than he was in building me.

Steering your offspring safely through dependent childhood and vulnerable puberty into self-sustaining adulthood is cause for celebration. When our children are young we command and direct them for their own good: 'Don't touch that, it's hot!' 'Keep away from the edge of the pond!' 'Don't put that fork in the socket!' It's a wonder they survive at all. When they are teenagers we try to steer them away from drugs and alcohol and underage sex by imposing rules and curfews. Billy directed his business in a similarly autocratic and protective way. 'I want that order out by Tuesday and there will be no excuses.' 'You made the mistake. You get on the phone to Aberdeen and sort it out.' 'People are wasting money. I want to see every order and expenditure by first thing tomorrow morning.' My father's position as head of the company was rarely in question. There were only a few occasions when he was challenged and even fewer when someone other than Billy emerged better out of the confrontation. An employee once went into Bill's office, saying, 'I've been here six months. I think I'm doing a good job. How about you give me a pay raise?' My father's response was: 'You've been here five. I don't agree. How about you leave?'

Mistakes are made and lessons learned—in business as in childhood. With luck, they're small and can be laughed about afterward. Like the time that Billy had just given up smoking. After five days' abstinence, he was experiencing what the Royal College of Physicians would describe as a withdrawal phenomenon, of which 'the most prominent symptoms are anxiety, restlessness, poor concentration and irritability or aggression'. When Andrew disagreed with the chairman on what might normally be regarded as a minor technicality, not worthy of so much as an arching of the eyebrows, Andrew was told, more in bleeps than in words, to collect his pink slip. Andrew gathered up his notepad and papers and, without another word, stormed out of the office . . . but only as far as the fuel forecourt next door, where he bought a pack of Rothmans. He marched right back in and slammed them on the desk. With the words: 'Don't be daft. For God's sake start smoking again and give us all some peace!' he reinstated himself, averting what would have been a big loss for the company.

And in 1988, it was John Addison's responsibility to install the scarily expensive new plant that converts the production of soda lime from a batch process into a continuous flow. There were predictable headaches over delivery of the components, timing and the quality of material produced. Day after day the product was not up to spec and was thrown out. John had been problemsolving up to fifteen hours a day. For weeks, and with uncharacteristic restraint, my father stood back and let his trusted manager get on with the job. Then he snapped and sent him an ornery memo about something unrelated and so trivial neither of them can to this day remember what it said.

John marched straight into Billy's office. 'I have thirty years' experience in the business. I'm a middle-aged grandfather with three grandchildren and neither you nor anyone else is going to treat me like this,' he shouted. 'I won't tolerate it. I'll walk out.' My father said nothing, but the next day he handed John an envelope addressed 'To a middle-aged grandfather'. In it were two first-class tickets to Portugal so that John and his wife could take a holiday. It was the closest to an apology as anyone ever got from him.

Those adolescent outbursts were little more than throwing forks against the wall and those days are over now. Of course there have been difficult times: the day Peter Addison, John's son, burned his hands and face in a chemical accident (from which he subsequently recovered); the night a fire destroyed the newly installed oven; and, worst of all, the early-morning call from Joyce Cheek to say that her husband Bill, the foreman, had tragically, unexpectedly, died of a heart attack; this was the only time in my mother's memory that she ever saw my father cry. But those were the low points. The company my father built has grown up and it's time to celebrate.

Without the benefit of a genetic blueprint, Billy more than built and nurtured Molecular Products. By its Silver Jubilee the expanded business, with its subsidiary in the United States, has become an extension of him; family, intimate, shaped by his thinking, the product of his soul. He built its structure, function and personality; he fought for the income that gave it life, chose its employees, directly and indirectly molded its character. He closely monitored its well-being and spared nothing. And now his youngest offspring is rational, permanently solvent with a turnover well into the millions, a healthy order book and a stable workforce of sixty. Twenty-five years in business was always going to merit celebration. And I knew, from experience, that my father did celebrations with style.

I remembered how my twenty-first birthday had fallen on a day when Billy was away, in Singapore, on business. To ease my disappointment he said he would try to get back for my party the following weekend, but he couldn't promise. By the morning of the big event there had been no news from him, so I went off to work as usual. I had progressed from shelf-stacking to helping out in a more upmarket boutique to supplement my student grant. Anna, the chic Italian owner of Fantasia, agreed that I could leave at lunchtime to finalize the party arrangements.

On the stroke of one, in strode my father. A man in his forties walking purposefully into a ladies' clothes shop immediately attracts attention, especially if manhandling a cardboard container the size of an ironing board. Customers and staff gathered round and gasped. It was packed with fresh, vibrant orchids, and the rare magenta, white and tiger-striped blooms were more exciting than the latest fashions from the Paris catwalk. There was no need to introduce him to the rest of the staff. The international bearer of exotic gifts had already spoken to Anna days earlier and my pals were all complicit in the surprise.

For the Molecular Products birthday party, plans have been no less intricate. The noise from the huge green marquee is a deafening hum. Andrew is on duty at the site entrance, welcoming the guests—of which there have been many: suppliers, customers, dignitaries and local residents; today the company is open to the entire town. Class three from Thaxted Primary are currently touring the labs and production plant while John Addison is rounding up the next group and my mother is handing out tea to a couple of the elderly neighbors. My brother looks weary. Maybe Andrew's tiring of being the host, or is he still the host's son? I'm not really quite sure.

'Sorry I'm a bit late. How's it going? How's Billy holding up?' I ask, fleetingly kissing him on the cheek and scanning past the people I don't recognize, looking for that one familiar face.

'He's all right, but I'm knackered!' My poor brother has hardly slept for the last two nights, getting everything organized. There is no end to the promotional details that my father has dreamed up for others to sort out. There's the new website, balloons for the children, sweatshirts for all the staff, flowers in the company colors, felt-tipped pens emblazoned with the orange hexagon-shaped company logo, and there are even two commercial divers demonstrating their re-breathing apparatus in a five-meter tank of water outside the research building.

'He's over there with the bank manager.'

I can't tell which jowls and paunch he's pointing to, but no matter—I am so used to checking on my father that I could pick out his pinstriped frame blindfolded. He is generally the person at the head of a group anyway. As Brian Colley, one of his longest-standing business colleagues, once told me: 'When your father walks into a room everyone wants to talk to him. You know he's lived and has something interesting to say before he even opens his mouth.' The closer I get to the table where he is holding court, the worse he looks. My heart turns over in terror. His pallor is terrible; his face is green. What does it mean? Is it an infection, a blood problem? What's happening? His skin has the dull, lackluster tone of an Australian toad. And yet he is full of beans and, quite delighted to see me, beckons me to pull up a chair.

'This is my daughter, Ruth,' he announces to the group. 'She's the goodlooking one in the family—obviously takes after her father.' I sit down, pleased that I made the effort to put on a suit and some mascara. The mood is convivial. Half-filled wineglasses and the lingering smell of olive oil and chocolate are all that remains of the catered lunch. I wish I'd been able to get here in time to enjoy it but I had immovable meetings of my own to attend.

Jim Beard, the accountant, is reminiscing about the early days before the company reached its current solid financial position. 'When I went through the books with Bill once,' said Jim, 'I was explaining that quite a lot of the expenditure was down to general expenses and do you know what Bill said?' The accountant looks to those around him, who, on cue, appropriately shake their heads, and then to my father, whose features are softening in readiness for the opportunity that Jim doesn't even know he's offering.

'I said: "I didn't know we were paying any army officers!"' Billy interjects, grinning.

My father's tongue can have the speed and accuracy of a chameleon catching a fly. Jim fades into the chorus, his punch line trumped, while his peers laugh so hard that a couple of coffee cups on the table rattle and the orange lily centerpiece collapses in the fallout. Yes, Billy is in good form, although his lively bonhomie contradicts his wan appearance. I glance back to Andrew, acknowledging that all is well, and—that's odd—he looks a bit peely-wally too. Actually, everyone seems off color. Then it dawns on me. The assembled group is suffering from the same complaint; sunlight through a green can-

vas turns everyone into amphibians. In this case, I have been worry-
ing for nothing. I let out a small sigh of relief and prepare to be more
upbeat. After all, this is a day for celebration.

With the flow of lighthearted conversation at the opposite side
of the table I can't stop myself. I whisper to my father: 'It all seems to
be going very well. How are you feeling?'

'Very tired,' he says. 'I'll have to go for a lie-down or I won't be
able to last until dinner. Hold the fort for me, would you?'

I should have anticipated that he wouldn't manage a whole day
of repartee and feather-spreading. He's been on the go all day and
has been entertaining clients and customers all morning, before I
arrived.

After he has slipped out I respond to comments concerning his
health confidently but dismissively. 'Yes, it was quite a shock,' I say.
'But he is well on the way to recovery.' I reassure so many people that
I almost start to believe my own propaganda, even though I know
that his best business suit has been taken in to fit and he's wearing an
old pair of lace-ups from the back of his wardrobe because his Bally
slip-ons will no longer stay put on his slimmer feet.

My father hasn't seen many of the people here since Ian
and Andrew took control of the company. Billy was anticipating lay-
ing his success open for all to see, and proudly introducing his sons
as the men who will now carry on the business he built. He wanted
to hear people say how well he had done, what a success he had
made of his family, of his life. I can hear him thinking: Not bad for a
wee boy from Blairgowrie!

That feeling is undoubtedly here, in the room. Perhaps a little
understated, a little short of the hearty backslapping he might have
wished for. And it's not that his former colleagues aren't impressed,
that his suppliers don't recognize the quality of the company they
deal with, nor that the local member of Parliament, Alan Hazelhurst,
doesn't value the contribution the company has made to the local
economy. If there is any restraint on the day's spirit of jubilation,

Billy himself has cast the shadow. Many of those present haven't seen him since he was ill. The room is buzzing with questions about his health, with different perspectives on what might have been, and everyone wants to recount their own version. 'He caught it on the golf course, you know,' said Jim Beard to one of the dignitaries. 'He was in the hospital for weeks. We went to see him and he looks a hell of a lot better now than he did then,' said a visitor. 'It's a good job Bill had already handed on the reins to Ian and Andrew. Lord only knows how the company would have managed otherwise,' said another.

Today was designed to be a celebration of Molecular Products' emergence into adulthood. I can tell from the stack of empty wine bottles, the free-floating balloons and the overflowing parking lot that, as my father would say in one of his more autocratic moods, this goal has been accomplished. But the day has become more than the celebration of one small chemical company's growth and endurance.

Today has become a celebration of my father's survival. But for the benefit of Intensive Care medicine, I shudder to think what it would have been.

9

Stress

THERE IS AS MUCH CHANCE OF STOPPING MY FATHER'S TRIP TO FLORIDA AS
there is of preventing the autumn leaves from turning. The very day
that Dr. Marcus declares him fit for travel he books two round-trip
tickets to Sarasota, leaving just days after the company's Silver Jubilee and returning the second week in March, in time for their annual
golfing trip to Portugal.

My mother is worried; she knows she can't stop him but what
will they do if he falls ill three thousand miles from his hospital
notes, their friendly general practitioner and the rest of the family?
Where would they find a hematologist as expert as Robert Marcus?

She might be worried but I am not. In the unaired debate, I side
with my father for many reasons. First, visiting Florida has been one
of his three wishes and it's a daughter's duty to see her father's wishes
fulfilled. Second, his chances of survival are better in Sarasota. Medi-

cal statistics say that CLL sufferers don't die of leukemia; they die of something else. My father is much more likely to pick up an infection in the damp, chilly British winter than in the Sarasota sun. Third, if five-year-survival statistics are an appropriate index, American medicine is superior to the British National Health Service for almost any serious illness you care to choose. Cancer is managed much better in the colonies.

Dr. Marcus has written an open letter to whichever doctor takes Billy on. 'The outlook is variable,' it says. Sounds like a weather forecaster, I think—and how often do *they* get it right? It could be possible to be reassured by the further comment that his patient's CLL is responding well to treatment and that 'he could enjoy a period of sustained remission for many years', but the hematologist's carefully crafted words don't fool me. I've seen enough clinicians' notes at St. Mary's Hospital to read between the lines. The term 'variable' brings little comfort. Truth is, he could go at any time.

I understand why he wants to visit Sarasota. Everything about the place oozes warmth. The sun shines predictably and ceaselessly. Palm trees sway in the breeze and gardens are filled with shrubs built from the same blueprint, held up to a fairground mirror. Great swathes of tropical hibiscus and mauve bougainvillea punctuate the landscape. The houses in the little cul-de-sac where my parents have their second home are painted soft pink like a baby's skin and their doors are the color of old-fashioned sticking plaster. In my father's view, the birdsong is louder, the air sweeter, the flowers more beautiful, the climate cleaner, the future rosier. Even the beds are more comfortable in Sarasota, he said. If life can be restored anywhere it will be here, in the deliciously stress-free land of the able geriatric, where doctors are used to treating disorders of the aging population. My parents are relative youngsters on Florida's Gulf Coast, where the citizens seem older but happy, tanned and medically well-cared for. Yes, Billy will be fine.

The stress that gradually builds up in Saffron Walden has always dissipated in Sarasota; that's one reason he bought the house in the first place. Billy had spent many holidays playing golf in Florida, from Lake Nona in Orlando to the Oaks just south of Sarasota. His intention had been to let go gradually of the day-to-day running of the business, leaving Ian and Andrew to manage Molecular Products while he and my mother moved gracefully toward retirement, ultimately splitting their year between Sarasota and Saffron Walden. My mother's teacher's pension provided the down payment on 304 Woods Point Road, a place where they could seriously wind down. The two-bedroom duplex with office, lanai and deck offered the lifestyle he had promised himself, the luxury they both deserved for a lifetime's hard work.

Before his illness, anyone would believe that Billy's master plan was coming together. In England, his out-of-season tan and the increasing frequency with which he checked his calendar and said with mock disappointment, 'No, I'm sorry, I can't make it—it looks like I'll be in Florida,' would give the impression that life could not be better. In Sarasota, their home from home, with unlimited pool access, four sets of golf clubs and a bike to ride to Nokomis beach, relaxation and well-being were guaranteed. I've even noticed myself how I walk a little taller along Casey Key, how my spine tangibly decompresses. I feel lighter, fitter and happier as the weight of work burns off. Out there, I have even been seen jogging. For, when you're in good spirits, you're in good health.

Gone are the days when scientists like Descartes and his like-minded dualists saw the mind and the body as separate entities; we now know that they're in constant dialogue. Separating physiology and neurology was normal in the early 1600s but, over the centuries, science has brought our understanding to almost the opposite of that belief; the brain so fully controls human physiology through well-mapped, closely coupled circuits that mind and body collapse into one. This is particularly true where stress is concerned; its effects on brain and body can't be separated.

The body's response to stress depends on three organs: the hypothalamus, the pituitary and the adrenal glands. The hypothalamus, the ancient brain's core, has long-term control over blood pressure, body temperature, fluid balance and body weight. It regulates digestion and reproduction and manages the body's response to external information. It coordinates all this through the release of hormones. The pituitary is a pea-sized structure tethered to the base of the brain, a 'master gland', linked to the hypothalamus by a neuronal motorway. Its job is to respond to changing circumstances by releasing any one of its dozen or so hormones on the hypothalamus's command. The adrenal glands are two golden triangles of tissue, like soggy tortellini, sitting one on top of each kidney (I know this because I once saw a surgeon pluck one from a corpse at St. Mary's Hospital), that release different hormones of their own. Together this triumvirate, the body's integrated system for managing stress, is known as the hypothalamic-pituitary-adrenal axis or, mercifully, the HPA axis for short.

Acute stress—the seven a.m. telephone call, the sight of our own blood or the sound of our children screaming—immediately activates the 'fight or flight' response. Physiologically this is apparent in our hearts beating faster, blood rushing to large muscles which can tighten or tremble, respiration increasing, pupils dilating, and we can sweat profusely too. This spectrum of reactions is caused by the neuronally stimulated release of adrenaline from the adrenals. It is the fastest route to a response; the body's 911 call.

Chronic stress, on the other hand—the death of a loved one, financial worries, upcoming exams—activates the HPA axis in a more sustained way, setting in motion a chain of hormonal commands that have more lasting effects on the body. The hormones respond more slowly; their role is to monitor external pressures, act as social barometers and manage the body's responses accordingly. The hypothalamus releases corticotrophin releasing factor (CRF, of which more later), which tells the pituitary to release ACTH (adrenocorticotrophin), which in turn tells the adrenal gland to release a variety of hormones of which cortisol is the predominant one. Cortisol helps to regulate blood pressure, the immune system, blood insulin levels—all of which are components of the body's response to stress. High cortisol is the body's classic hallmark of chronic stress, reflecting a 'state of high alert'. But in the sunny, calm weather of Sarasota, cortisol should be able to stand down from its duties and slip back to the deepest recesses of the adrenal gland.

My father is outside organizing storm shutters when we arrive. I hug him as warmly as is seemly while he continues discussing the merits of automatic-locking mechanisms with the shutter salesman. I scan him surreptitiously. He doesn't look so different from the last time I saw him. Not fatter, not stronger, and, disappointingly, not happier either. I go inside to ask further details from my mother.

'How was the trip over?' I ask her.

'Exhausting. We had to wait in line at immigration for over an hour. It was the worst I've ever known. I thought your dad was going to pass out but he wouldn't give in and let me get him a wheelchair.'

'Is he doing his exercises?'

'Mmm . . . not really, but I do think he's lost some of his pot-belly.'

I fill two glasses with iced water from the dispenser on the fridge door and take them out to my father as an excuse for a closer look. The salesman has left and Billy is lying on a recliner, admiring the orchids hanging in wooden baskets round the deck. *Cattleya* Wilder Point, a large, yellow-centered flower, its petals frilled like the shirt cuffs of a Renaissance cavalier, is in full bloom. Yes, perhaps he is a little trimmer. He is wearing shorts and that makes him look fitter, although his legs look thin, but thinner than before? I have paid little attention to the musculature of his calves; it's not the sort of thing a daughter does. Strong perfume from the orchid fills the air; sprays from another phalaenopsis hang down over us. No matter—

I'm here now. I'll implement a strict regime of exercise and stress-free living.

We both know that he needs to persevere in building up his strength. For my father, strength means physical strength. A full eighteen holes requires more muscle, more power in his arms and lungs. For me, exercise means mental strength: feeling better is just as important as being better.

It seems almost second nature to know that exercise is good for you. Runners have long proclaimed its benefits. Serious exercise releases endorphins (the brain's *endo*genous mo*rphin*e), the pleasure-inducing opiates, into the blood. Runners attain levels fivefold higher after exercise, which may explain the so-called 'exercise euphoria' that dedicated fitness enthusiasts experience. Endorphins are just one of the pituitary's many hormones, produced from the same precursor as ACTH.

But, for me, the most exciting thing about exercise is that it is good for the brain. It increases neurogenesis—the number of new neurons that are born. Fred Gage and his colleagues from the Salk Institute gave laboratory rats unlimited access to an exercise wheel. On average they chalked up three miles a day, a distance that would put couch potatoes and even members of the Saffron Walden Striders to shame. Closer examination of the rodent brains showed that the birth of new neurons in their hippocampi had more than doubled. Furthermore, in the water-maze memory test, where rats have to remember the location of a platform hidden just under the surface in a pool of opaque fluid, well-exercised rats could remember the platform's location better than their sedentary siblings. Long-term potentiation, the electrical process that strengthens synapses and underlies memory, was also increased, suggesting that regular exercise might improve cognitive performance too.

Starting tomorrow, a whole new regime will begin. There will be swimming and relaxing and recovery.

With rolled-up towels under our arms my father and I set off early to swim before the sun, or the rest of the family, is fully awake. Arm in arm, we pass the pink houses, the pine trees, whose offerings litter the tarmac drive of number sixty-four, and walk along the white crushed-shell path. The pool directly overlooks a creek, and in the hazy morning light, with cicadas humming and fluorescent blue dragonflies darting occasionally across the water, you could believe that the one flows into the other. That first length, heading away from the pool house, is tough. The water isn't warm yet but we try not to care. We are here to exercise, to recuperate. We have no time for namby-pamby whining about the temperature. The surrounding mangroves and tall white-stemmed pines provide security, the sensation of being pampered in a warmer paradise. And when rays from the rising sun reach the far end of the pool we swim into its light, soak up the pleasure of heat on our faces and feel we can swim on forever.

Forever, in my case, is actually twenty lengths. On a normal day Billy does six and on a good day eight. When he reaches ten I want to believe he is immortal. As I shuttle backward and forward he exercises in the water: knee bends to strengthen those slim-line calves and leg swings to improve flexibility.

'Good for the golf,' he says, grimacing with pain.

Bending with outstretched arms exposes the lemon-sized lumps in his armpits, lymph nodes stuffed with impotent cells, the tombstones of leukemia, draining his body of its resources and contributing nothing. God, how I despise them!

Showered and dressed, we emerge on the white shell road like Dorothy and the Tin Man. There is a place on the path where the sun blasts through the trees, square onto our backs. I come to expect it and, were I able to sing, 'Somewhere Over the Rainbow' would not be out of place. At home, the coffeemaker is ready with endless mugs of 'choc full o' nuts' that gain in strength as the morning unfolds.

'We're thinking of going down the coast a bit today. Dad, should we drive or take a boat?' I ask.

'You can get a boat from round by the marina but I think it only goes as far as Venice. If you want to go to Captiva you'll have to get a bigger boat up in Sarasota Bay.'

Captiva? No, I don't really want to go to Captiva Island again. My father took us all there for his sixtieth birthday, another of his family celebrations for the whole McKernan clan. I knew it would be fabulous: wide sandy beaches, bikes, sailboats, nature reserves to educate the youngsters (all eight of them, and the oldest was only nine at the time). There were also fire ants, *solenopsis invicta*. And no one escaped the pain of Captiva's sting.

That holiday, four years ago in 1995, should have been the pinnacle of Billy's success, the family get-together of a lifetime, the special shareholders' meeting where the new MD and sales manager were going to impress the rest of us with their plans for the whole family business's future. On the surface, Billy's plan to loosen the reins and spend more time in Florida seemed to be coming together. For years, he had been personally grooming Ian and Andrew to take over the business. Ian had secured his first deal solo, an accomplishment that Billy had made little comment on at work, but when the fledgling magnate got home he found a case of wine on his doorstep with a note saying 'Congratulations—from the Chairman to the MD'. Andrew earned his stripes after a management training course in which, as the least qualified among a group of Oxbridge graduates and seasoned directors, he led his team to victory. Billy admired the way he handled colleagues who didn't agree with his strategy. In a competition to transport bricks across a river and build a wall he simply ignored them and single-handedly completed the task himself while other members of his team were still taking an inventory of building materials.

My brothers flipped through their presentations. They described plans for restructuring the existing business into a holding company and subsidiaries. They proposed erecting new buildings and expanding into new product areas. On the surface it sounded good and looked good. Backs were patted, corks popped, sighs of relief were heaved.

My father knew the statistics on family businesses. The odds on their survival from one generation to the next are no better than fifty-fifty. He was determined that his offspring would not be in the losing half. Billy had given great consideration to how and when he would hand over the business. Like King Lear, his intention was 'to prevent further strife'. However, while the monarch of Shakespearean tragedy came up with the idea of dividing his kingdom according to how much his daughters loved him, thankfully my father took a more financially astute approach. Just as well, for if he had asked us to quantify our love, I may well have answered like Cordelia, that I loved my father 'according to my bond—no more, nor less'. And as a scientist, I know the sheer magnitude of that bond.

My father's approach was driven by business, not love. First, he had taken legal advice on when to divest shares, how to avoid Capital Gains Tax. Six years earlier, he had written us all a letter, as he tended to do for matters he wanted taken seriously. 'As the shares continue to appreciate, so does the tax problem which is stacking up for all four of you, if your mother and I die without making arrangements to dispose of our shares early enough,' he wrote. He proposed gifting the majority of the shares to Ian and Andrew, 'because MP is their future and essentially will become their business'. A nest egg of 50,000 shares each was kept back for my parents, should they want to cash it in later, and a lesser share holding was designated for my older brother, Stuart, and me, compensated for by a larger share of my parents' properties.

Second, he had bought the house in Sarasota, somewhere he could retreat to while his sons gained confidence running the business but yet be accessible at the end of the phone or fax to give them the advice and encouragement he expected them to seek. Third, he put in place a formal succession plan, signed Ian up for the Institute of Directors and sent him on training courses to learn the role of

director. He had Andrew shadow John Addison as production manager for a couple of years before moving John to head up Operations
and leaving Andrew to run the department.

My father tried so hard to be fair, ensuring that what we received
was of equal monetary value, taking into account the contribution
that Ian and Andrew had already made to the business. It should all
have worked out neatly. Billy no doubt believed that he was going to
relinquish control. But when it came to putting his plan into practice, to standing back and watching others—even his sons—making
mistakes with his money, failing to make the same decisions he would
have made, he suffered from incredible stress. Like Lear, the process
of passing his estate to his children would 'upon the rack of this
tough world stretch him out longer'.

What my father gave away with his right hand he held onto with
his left. The travel plans Ian made were changed by Billy and then
changed back again by Ian. He told a long-standing customer that
Andrew was now handling their business and then phoned him up
to discuss it anyway. Billy was as suited to the role of bystander as he
was for a lead in the Bolshoi Ballet. Molecular Products had been
built without delegation or discussion. He didn't have the cerebral
circuitry for such processes. He didn't have the genes.

The handover from father to sons had more bumps in it than a
phrenologist's handbook and worse was yet to come. In Captiva Island, behind the PowerPoint presentation, congratulatory champagne cocktails and the world's best Caesar salad lay a different
reality. Family businesses hold together about as well as families do.
My mother and the Queen had much in common in 1995: a dress
by Frank Usher, expensive shoes in exotic leathers, and another child's
marriage heading for the rocks. As the holiday wore on, it became
clear that Ian and his wife were moving toward divorce. Mealtimes
were briefer, attendance incomplete and laughter rationed. Family
discussions focused almost obsessively on our small children, the only
safe topic of conversation. Such vagaries of life had not been factored

into my father's calculations. Put simply, if my sister-in-law claimed half of her husband's total shareholding in MP, the entire future of the company was at risk. Captiva Island? A picture of paradise etched in formic acid! I'd rather not go there again.

'We'll just go along the inland waterway. A couple of hours is probably all the kids can cope with anyway,' I say.

'But we don't want to go on a boat. We want to go to Halloween,' chimes in Francesca, having seen the pumpkin-colored signs announcing the annual Trick or Treat parade through Sarasota Square Mall.

'Take us to Halloween, Granddad!' and she climbs up on the chair beside him. I seize the moment, pick up the camera from the glass table and carefully frame the photo with their heads close together, top left, and their feet on the stool, bottom right. It isn't often Billy lets the children cuddle up to him like this. How many more chances will there be to capture the two of them together?

'Of course we can,' he replies. 'Have you got a costume?'

We are ill-equipped for fancy dress. I make a cursory attempt at a cloak and hat with some old clothes of my mother's and a bit of cardboard that I found in the cupboard and do the best I can with the standard contents of a makeup bag to turn Francesca's cheery round face into a witchlike horror and I paint Adam's to look like Dracula, but the homogeneous black and green smudge stands no comparison with every other professionally face-painted, fully decked out, junior superhero.

At the mall, shoppers file past empty shops to receive candy in the doorways from seasonally dressed shopkeepers. Gobstoppers, lollipops, chewing gum and, more commercially, money-off vouchers are stuffed into bags or buckets to the chant of 'trick or tree-eat'. Determined to enter into the spirit of the day, Billy buys two large bags of Hershey bars. But unlike the locals, who have grown up with the custom of giving away candy, the company chairman can't quite manage to offer sweets to small children he doesn't know. So he sits

on a bench in the middle of the mall and watches the crawling conga
of fairies, furry animals and grown men with chopper-through-the-
head hats.

Adam tires of traipsing from shop to shop and sits down beside
his granddad.

'Gggrrrrr!' yells Billy and raises his arms above his head in
an attempt to be scary. The hollow facial features help but he is less
forbidding than many other characters in the mall. Adam jumps a
bit. He smiles at the more obvious ghouls yet he shies away from one
tall, motionless, skull-faced character. The reaper's voluminous cloak
reveals yards of ankle chains dragging on the floor. A dangerously
convincing scythe supports the man's skeletal arms. Adam can't stop
staring; his attention is so completely focused that when his grandfather growls a second time he doesn't even flinch.

We learn what is stressful and, similarly, we learn what isn't. Repeated exposure allows us to adapt so that the original stress no longer
matters. Without such a system the world would be rather shorter of
airline passengers and few would gain the confidence to pass their
driving test. Cortisol release is carefully controlled by the HPA axis
with strong inputs from the hippocampus and amygdala allowing us
to rationalize what is happening, giving us some control over stress.

Control is all-important in stress; those in control suffer less.
Two huge studies, one of 17,000 British civil servants, known as the
Whitehall II Research project, and the other of a million employees
at the American Bell Telephone Company, both showed the same
thing: the factor that most consistently matches stress-related illness
isn't cholesterol level, body weight, smoking or blood pressure. It's
the job you do, your employment grade. The more junior you are,
the grimmer your future health. In the civil service as well as the
telephone company, senior administrators at the top of the tree had
only a quarter of the mortality risk compared with junior clerks. In
the Whitehall study, the worst sickness rates were among those who
rated their jobs low on control. The Donald Trumps of this world

don't get ulcers, they give them. Perhaps executives avoid stress by transferring it to others. The lowest on the totem pole, the Dilberts of office life down there in the dingiest cubicles, have no one to transfer their stress to, and their outlook is concomitantly bleaker. As for the long-term unemployed, their statistics are the worst of the lot. When it comes to stress-related disorders, having no job is worse than having a low job.

Social studies on primates concur. Some of the world's leading primatologists, including Robert Sapolsky from Stanford University and Barry Keverne from the University of Cambridge, reviewed a collection of studies asking why cortisol rises in primates. Squirrel monkeys and baboons, it seems, do not differ that much from civil servants: rank matters to them too. In primate colonies, animals lower down the pecking order, the withdrawn and weak, have higher cortisol levels. More detailed analysis says that it's not rank *per se* that matters. It's rather what happens as a consequence of it. For example, if the alpha male dies and the established primate hierarchy dissolves, subordinate animals jockey to reposition themselves and their cortisol climbs. Levels rise higher still if inequalities are maintained by hostility.

By late 1995, the tension between father and son was at its peak. Rather than follow my father's views, my brothers developed their own. Ian felt that the business should be less ambitious in its development plans and much more prudent with the cash it had. While Billy was improving his handicap in the Sarasota sunshine, Ian implemented a swift and fairly draconian set of measures to cut expenditure, de-stock and close a couple of the research projects, countermanding some of my father's earlier plans. He outlined his thinking in a memo to my parents, copied to Andrew. Furthermore, he asked my parents to be more accountable for their own expenditure.

This went down predictably badly. In Billy's mind, it was still his company, and by return of post, in frenetic spiky script, came a letter that must have had every molecule of cortisol in his body out on patrol.

'*I am responding briefly to your ill-advised letter of Feb. 6th, if only to acknowledge receipt and establish immediately my complete disagreement with your analysis of the business situation at MP.*'

The letter went on to suggest that the company's poor performance was due to Ian's lack of direction, drive and presence, not his, and alluding to his marital circumstances it finished: '*I urge you to look to put your own house in order before you suggest your mother and I are stifling the company's growth by stripping it of cash . . . God help MP if this is how you conduct yourself in personal matters there! Dad.*'

If only Billy had been able to manage the stress better. If only he hadn't reacted so badly to the things that Ian wrote or did. If Billy invested less or cared less, perhaps he could have got through this with less damage to his health. For it is now without doubt that chronic stress is about as good for you as a pack of cigarettes a day.

Stress is bad for the heart, bad for the immune system and bad for the brain. Bruce McEwen from Rockefeller University in New York was one of the first to show that chronic emotional stress produces physical changes in brain structure. He studied tree shrews that were deliberately housed in pairs where one was a dominant aggressive animal, the other a stressed subordinate. The dominant animal regularly bullied the subordinate and when both animals' brains were examined the bullied subordinate had fewer dendritic spines, a shrunken hippocampus and fewer new neurons compared with its bullying roommate.

Cortisol is the culprit. The McEwen lab reproduced the effect by implanting animals with pellets secreting high doses of the rodent equivalent of cortisol. When cortisol is high, neurogenesis decreases, they concluded. Further studies show that the reverse is also true: when cortisol is low more new neurons are born. Somehow, this has to be linked to depression. Chronic stress causes depression—that has been observed in many studies, and anatomical information shows that severely depressed subjects have slightly smaller, less active hippocampi. Furthermore, as Ron Duman, the Yale Professor of Psychiatry, found, rats treated with Prozac for three weeks have 70 percent more new neurons in their hippocampi. The new knowledge that neurons are continually being born and replaced is leading scientists to reconsider what we know about depression and how antidepressant drugs work. Is a stress-induced elevation in cortisol (and other hormonal components of the HPA axis) enough to cause depression? Is this yet another detrimental consequence of chronically being on red alert?

I put the overflowing baskets of treats onto the glass table. Billy is looking through his binoculars at the birds in the mangrove just beyond the garden boundary.

'So, how do you think you're progressing?' I confidently ask my father, prepared for a positive reply, perhaps even anticipating a reward for my dedication. Yes, thanks to me, he'll say, thanks to his dutiful daughter's vigilance, he is definitely improving, hasn't felt so good in months.

'It is still taking so much longer than I thought it would,' he flatly replies. Then he adds that he isn't putting on any weight; in fact he is steadily losing it. 'There are times when I feel almost depressed,' he says.

My heart sinks. I've never heard my father use that word before. Even in the hospital, when all his energy was going into physical rebuilding with little left for mental restructuring, when there was much to be depressed about, he never used that word. Even when I found him sitting in the big grey leather chair, at home in his study three years earlier, uncharacteristically staring into space and looking depressed, he said he was fine. In retrospect, that would have been at about the same time that his CLL was diagnosed, although I didn't know it at the time. He hadn't been himself that day. He didn't want coffee, he didn't seem to want to talk, but he hadn't claimed to be depressed then, either. Now the look on my mother's face says that the weight loss is news to her, too. Sarasota looked so much better in the wrapper.

Losing weight? How can he be losing weight? I must try harder to build him up. Forget the salads. Banish celery from his Bloody Marys; it takes more energy to eat a stick than can be gained from its largely cellulose content. For lunch, I make bacon and avocado sandwiches. I leave a thick ribbon of fat on the bacon and slap on spoonfuls of mayonnaise too. I dream up menus for the rest of the week made from all the things I avoided as a weight-conscious teenager. I become more covert, sneaking protein powder into his milk, spreading butter extra thick, offering ice cream when he asks for a glass of water.

Is he losing weight because he's depressed or is he depressed because he's losing weight? Or is some other factor responsible for both depression and weight loss? The mind is in constant balance with the body. Cortisol sees to it that the two are in synchrony. When cortisol is high the message to the body is 'free up capital', de-stock, call in outstanding debts—you might need it. Don't invest in ambitious new acquisitions. This is not the right time for long-term investments of energy like having a baby or studying for a degree in higher mathematics. Cortisol is telling the body to mobilize resources, to be on alert, and energy is released as glucose, or, preferentially, stored in more accessible fat, rather than in muscle or inaccessible bone. Only when the body is thriving is it time to build infrastructure of muscle and bone and to train the immune system and the brain.

Being ever ready is hugely beneficial in the short term, but if you carry on like that indefinitely there will come a time when chronic underfunding of the infrastructure begins to tell. The physiological consequences of chronic stress—prolonged elevated cortisol—also include reduced bone formation (and consequent osteoporosis), disorders of the liver, the pancreas (including type II diabetes), the cardiovascular system, leading to potential heart attack and to ulcers, to arrested ovulation in women, and to impotence in men and—most notable in my father's case—an impaired immune system.

Chronic stress damages the body just as much as underfunding and short-termism damages business. It's rather one-sided to think of the HPA axis as being controlled by the brain as the autocratic CEO in charge of hormonal release. Actually, there is no hierarchy. The HPA axis is a self-regulating system, like the free market economy. There is no single controller in charge. Rather there are times when you should spend, and times when you should save. Longevity comes to those who sense which is which.

Stafford Lightman, Professor of Neuroendocrinology at the University of Bristol, has explored the damaging effects of chronic stress on the immune system. No one would disagree that looking after someone with end-stage Alzheimer's disease is about as stressful an occupation as you can get. Stafford found that people who cared for sufferers of Alzheimer's disease for six months had higher levels of cortisol in their saliva than their peers. He subsequently studied their ability to generate antibodies to a routine influenza vaccination. Those with the highest cortisol levels had the weakest immune response; they even had fewer infection-fighting lymphocytes.

It's just a short step from an impaired immune system to cancer. I can't help but wonder whether the chronic stress of finding a way to hand over the business triggered Billy's CLL. Would that rogue cell have escaped a more robust immune system? I can never know. I try to tell myself that his illness could just as easily have been a consequence of whatever radioactive studies he volunteered for while a student in Birmingham. Maybe it's all predestined in the nature of his genes—or is it just plain bad luck?

But, no matter, on some days my father is in good spirits. When the phone rings he answers with a strong, booming voice. He sounds like a big, happy, recovering man. His strength is growing; must be the exercise and the high-fat diet. One evening, when he is feeling well, we go out for dinner to the Flying Bridge, just at the end of Woods Point Road, and I talk Billy into ordering his customary rare steak. He needs the protein. For once, he eats heartily, washing the traditionally accompanying soup and salad down with a Californian Merlot. We talk, we laugh, and after dinner we sit at the bar and order brandies until closing time. Billy is at his sharp-witted best, perched on a stool, spreading his big hand round the glass and taunting the barman for locking the cheap American bourbon away while disrespectfully leaving the real Scotch out on the shelf with the mixers and brand-label gin.

The exercise and stress-free regime must be working. We do whatever we can to encourage Billy to relax more. Gerry sets up access to British newspapers on the Internet. The five-hour time difference means that Billy can start scouring the next day's papers at 11 p.m. each evening for the daily crossword. He routinely starts with his long-standing companion, the *Telegraph*, but is soon tempted into infidelity with other broadsheets by their sheer ease of access. On a good day he prints them off in the order *Telegraph, Guardian, Independent* and *The Times*. On a perfect day all the crosswords are completed by the time Molecular Products would be closing for business back home. Oh yes, things are looking up.

My father finds a local hematologist as Dr. Marcus had advised. Nothing adverse is found in his routine checkup: lymphocyte number is decreasing, platelets and red cells are within the normal range. Chlorambucil is still effectively treating his CLL. It is good news. I brush his comments about feeling depressed and losing weight way back into the recesses of my mind. The medical reports are good. These are real facts I can trust, quantitative measures concerning the state of his condition that lie outside the grey zone of mood and personal opinion. I give the blood counts top billing—more important that his white cells decline. After all, inadequate white cell function is his primary underlying problem.

My father tries to ignore other minor ailments. He gets toothache, which isn't a surprise as it's been an intermittent problem since he first became ill. It is probably related to an infection in his gums. As the risk of infection is always a concern he visits a local dentist. The oversubscribed, under-resourced dental surgeries back home make the practice here look very slick. The dentist sees no reason to have any teeth out. He can perform eight root canals and save all his teeth, a remedy that the pliers-happy British wouldn't contemplate. More good news, and maybe when Billy feels a little stronger he will have them done. But for the moment he no longer eats ice cream (damn—that was a great source of calories), takes his milk out of the fridge an hour before he drinks it and stops putting ice in his 7-Up.

Odd things happen. He has abdominal pains—probably only indigestion—so his primary care physician recommends some innocuous tonic. Incidentally, the doctor has a new vaccination for pneumonia. In the United States, it's standard practice for over-sixties to receive it annually and Billy can ill afford to get pneumonia, so he takes the shot. Then he gets aches in his back. He's probably overstretched himself playing golf so he goes to a local masseuse, who finds residual evidence of his once broken shoulder but, reassuringly, nothing else.

The next time we get out of the pool his skin erupts in an itchy rash.

'Must be some new cleaning agent in the water,' he comments.

The hives reappear all over his body when he showers the next morning, only to fade an hour or two later. Even when he washes his hands his palms turn prickly and red.

Being a scientist is a bit like being a detective and you can't stop yourself from trying to assimilate what seems like unrelated information into one whole sensible story. That vaccination worries me and I find myself contemplating something I've recently learned at work. Sometimes, when drugs are metabolized by the liver, they can become highly reactive and irreversibly attached to blood or liver cells. Blood cells with foreign particles stuck to them are recognized as being foreign and stimulate an immune reaction—anything from a mild rash to a severe allergic reaction. Several drugs, including the Parkinson's treatment Tolcapone and the diabetes drug troglitazone, have been withdrawn for just this reason. I also know that vaccination deliberately boosts the immune system, preferably against the chosen enemy, pneumococcus in this case, but wouldn't a pepped-up immune system respond to other alien signals more strongly? Chlorambucil binds to DNA in the excess lymphocytes, marking them out for destruction. Could an allergic reaction to the drug be the cause of his rash? Could it precipitate an immune reaction? I look for publications on the combination of immunization and chemotherapy with similar drugs, but come up blank.

Well, it was only a hypothesis and, no matter, he is definitely in better shape. His white cell counts are going down. He's back in the land of happy hour and onion rings and sunsets. If only Dr. Marcus could be right and he might enjoy this remission for some years to come.

His increased strength encourages him to take more notice of what is happening at Molecular Products. After his second call today, Billy puts down the phone, pours a glass of milk and settles down to watch the birds at the feeder. 'Your brothers are working on a big contract with the American navy,' he says. 'If this comes off, Molecular Products will be secure for another few years. They're doing OK, you know,' he adds, looking down and picking a piece of fluff off his shirt. Compliments flow from his lips as easily as water runs uphill. And how unlike him to comment on my brothers' performance. Even during the difficult transition period, my father's almost Hippocratic loyalty to his offspring prevented him from ever making any public judgment. But I could sense the pain of handing the business over then as I can now and I don't need to be a hematologist to know that Billy could no more give up the company than he could stop making excess lymphocytes.

My father and my brothers have completely different approaches to work. Ian admits that, on occasions, he sometimes gave less than he could have done. He makes no excuses for it. He thought my father worked too hard and that life is for living, not just for work. Ian had a new wife, a new life to live. When my father was building up the business he was rarely off the premises and he so consistently popped in on Saturday mornings to go through the mail that members of his staff often took it as an opportunity to pin him down for a serious chat. On Saturdays Ian played rugby.

When my father finally announced in January 1998 that he would be retiring at the end of the financial year there were many of us who failed to believe it. Once that decision became public the whole character of the company changed. Inadequate sales staff came and went like England strikers. Company stalwarts retired. Billy's long-serving personal assistant (PA), Carole, left to take up a position with the local estate agent. 'She might at least have waited until after I'd gone,' Billy confided in me. Loyalty is a trait he valued above all other.

My father's official retirement dinner in the private rooms of The Starr in Dunmow went well, on the surface. He took one look at the Georgian silver decanter and glasses on the side table, walked over to them and announced smugly: 'That'll be my leaving present. Very nice. Thanks very much.' And he teasingly picked up the decanter and tucked it under his jacket.

Billy kept a smile on his face all evening. The bravest face said all the right things. There would be more trips to Sarasota, long golf holidays, cruises aplenty. The boys would now be keeping him in the manner to which he had become accustomed, to which he was entitled. He boasted about how he had kept them swathed in luxury well into their adult years and now it was their opportunity—nay, their filial duty—to treat their old man likewise.

His thank-you letter, dated April 6, 1998, said: 'While it is a little sad to hand over the reins of the Company after so many years, it is nice to know that the Chair is in such good hands and that the

Board and the workforce are of such a high caliber to ensure the continuing prosperity of the company. Enjoy yourselves!' The words sounded right. The time had come to relax, to hand the stress to others. It should have been the beginning of something better.

Three weeks later, the phone rang at seven o'clock on a Saturday morning.

10

Falling Leaves

Now look objectively. You have to
admit the cancer cell is beautiful.

. . . The lab technician
says, It has forgotten
how to die. But why remember? All it wants is more
amnesia. More life, and more abundantly. To take
more. To eat more. To replicate itself. To keep on
doing those things forever. Such desires
are not unknown. Look in the mirror.
From 'Cell' by Margaret Atwood

I LOOK BACK TO THE INFORMATION BOARD FOR THE UMPTEENTH TIME. The slats flicker round like a harshly pulled Venetian blind. Against DL012 it records 'Baggage in hall', the same three dull words that have been there since I arrived. I take the tatty piece of paper out of my pocket, unfold it and reread the messages to pass the time.

WMCK3244@AOL.com   Sent: Fri 11/20/1998 15.39
Re: Workout kids

Dear Ruthie
    Well we survived the introduction at the Y. For starters we spent 15 minutes on the walking track. I had a walk-through demonstration of 15 machines on each of which I carried out one exercise 12 times, then to the pool where I completed 6 laps.

5 minutes in the sauna then out through the shower to meet
mum who had walked another 24 laps of the track—2 miles in
total. You'll be pleased to know that no rash appeared either at
the Y or at home after showering or a bath.

The workout was much more comprehensive than your
schedule but the instructor wasn't so attractive and was called
BUTCH. Tomorrow will be harder—8 a.m. at the Y followed by
golf at 10.45.

If I survive you'll be the first to get a report.

LOVE, DAD

That message was sent barely four weeks ago. Now it's a few days
after Christmas and I haven't seen my father for a couple of months.
Frequent messages and phone calls have indicated that since then
my parents' attendance at the gym has dribbled to a standstill and
the length of my father's walks has grown terminally shorter. They
have been to the beach, though; my mother told me that on the
phone. Nokomis is my father's beach; our favorite beach. I can see
him wiggling his feet in the sand, sifting the crushed white shells
between his horny yellow toes. The sand at Nokomis is special: adolescent shards with plenty of rough edges left for the waves to polish
ever finer. You can still make out black fragments of fossilized sharks'
teeth and the iridescent pearly white of abalone. My father could
while away a good hour or two searching around for whole teeth and
watching the water fowl: seagulls that scramble to attack the flying
fish like a squadron of Second World War fighter pilots; pelicans
that serenely patrol the skies, surveying the lapping water before folding their wings and dropping like a stone on their prey; avocets and
other birds that I can't name but he would instantly recognize. I can
see him sitting there, collecting grains under his toenails until the
pressure of the shifting sand grows too uncomfortable and he drags
himself back to his feet. I can see the two of them walk along the
wooden boardwalk, as they have so many times, over the sea grass
and back to the car.

WMCK3244@AOL.com   Sent: Wed 12/02/1998 23.47
Re: Florida Update

Ruthie, I saw Garcia today and got the following results:
White blood cells down from 18.2 to 16.8 thou per microliter.
Red blood cells virtually unchanged at 4.42×10^6 per microliter.
Platelets unchanged at 195×10^3 per microliter.

Copy of results going to Marcus by mail.

Obviously I'm pleased with the figures and am continuing with
another month of Leukeran. Your Mum is in the garden and is
calling for my help so must go now. Please don't spend all your
free time in the office, get out and smell the flowers.

Cheers for now. See you soon. Dad.

His white blood cells were still decreasing. Good news. Those
excess lymphocytes were as useful as a growing pile of junk mail.
They had to go. Marked 'For destruction' by the anti-cancer drug
Leukeran, their death is assured.

And death is good for you. Well, at least in small doses. I know
because I wrote about it once for the *Independent* newspaper, about
programmed cell death, the way in which the cells in our bodies die
and how this important process benefits the life of the whole organism. Death is something to be welcomed.

In biology, there are only two ways to die. Everything reduces to
either catastrophe (necrosis) or controlled deconstruction
(apoptosis). Necrotic death is akin to blowing up a building with no
regard for the structure or its surroundings, whereas an apoptotic
death sees it carefully dismantled with its appliances and period features reused. Given a choice, cells don't choose catastrophe; that only
happens when there is insufficient time or resources to orchestrate a

controlled death and it only occurs in particular circumstances. When my father had septicemia, the tissue around the infected eye went black, characteristic of necrosis and a consequence of the speed with which the multiplying bacteria starved the surrounding cells. Similarly, when he was in Intensive Care his toes went black from oxygen starvation because his compromised circulation could not deliver enough of the critical nutrient to keep his extremities alive.

Cells die necrotically when they are bankrupt of energy. The universal currency for energy in biology is a molecule called ATP (adenosine triphosphate, for the biochemically inclined). In the financial world, the money we make is kept as coins and bills. In the cellular world the energy produced from oxygen and glucose is stored as ATP. Each cell in the body contains mitochondria—molecular-sized factories—that generate ATP for whatever is required. Growth, cell division and fighting infection are all energy-dependent processes. Tissues in the body continually use ATP and even maintaining the cell's structure and integrity takes energy. Without it cells don't just grind to a halt, they die as dramatically as a ham actor in his first spaghetti Western. Losing control, they swell until they burst, 'spilling their guts' into the surrounding tissue.

As one might expect, the preferred option would be for a dying cell to take itself apart carefully. It's a whole lot tidier, less wasteful of resources and leaves far less mess for the rest of the body to clear up. The systematic dismantling of unwanted cells involves special enzymes and subcellular machinery that have remained dormant until now. Putting it into action comes at a cost: it requires energy; it requires ATP.

WMCK3244@AOL.com   Sent: Mon 12/21/1998 17.02

Dear Ruthie

Can you remember—whoever meets us—to have a couple of cushions with them. Can you please try to make an appoint-

ment with Dr. Marcus for any day including and after Dec. 30th.
Flight details to follow.
Love Mum.

I try to read the labels on the passengers' suitcases as they rush
past. No sign of the flight from Atlanta yet. Perhaps they're taking
more time because Billy is in a wheelchair. What if something happened on the plane? Where's the nearest hospital? If we were at
Heathrow I would know my way to St. Mary's in Paddington. I
know where the Accident and Emergency entrance is—or at least
where it used to be. But where is the nearest hospital to Gatwick?
Croydon? Is there one in Redhill? How would I get there from
here . . . and in the rush hour too? Could Mum get in touch with us
if she needed to? Yes, my mobile phone is on. I'll just check it again.
Should I call Ian, just in case they have called him? They know I'm
coming to meet them. They know my number. This is a mental
rampage. A little common sense, please. Just wait.

The contents of my parents' e-mails have become briefer. Phone
calls have been less upbeat too, with conversations covering only the
arrangements that need to be made for their return. Still, the information from Dr. Garcia was positive. The treatment is working. Let's
hope Dr. Marcus will have more to say when we see him.

Actually, I spoke to Dr. Marcus last week. I relayed the results of
November's blood tests and he concurred that my father should be
heading nicely in the right direction. But my own observations during my stay in Sarasota didn't match that expectation. There were
oddities: Billy had taken to drinking warm milk and swallowing ant-
acids or laxatives like M&Ms.

Having read that the incidence of other kinds of cancer is increased in people with leukemia, I worried that he might have some
kind of gastrointestinal cancer. Dr. Marcus's advice was to refer Billy
back to Dr. Garcia if I was concerned. On reflection, I knew it
was the only advice he could give. Diagnosing and treating a man

three thousand miles away is as sensible as an over-the-phone appendectomy.

Cancerous mutation, viral infection, radiation and even DNA modification, as in my father's Leukeran treatment, all drive a cell into apoptosis or controlled destruction. The term (pronounced a-po-toe-sis, with the second 'p' silent) derives from the Greek for falling leaves and describes the way trees shed their leaves in autumn, conserving energy and providing protection from the ravages of winter. Although the principle was first observed in the plant kingdom, it soon became obvious that this mechanism is a fundamental pattern throughout biology. Once given signals to die, cells in the human body, just like those of the mighty oak, can also be economically dismantled without trauma to the surrounding structure. Biology reflects art. As W. H. Auden said: *What is death? A Life / Disintegrating into / Smaller simpler ones.*

Why would cells choose to die? The dispassionate dictum of Mr. Spock from *Star Trek*, that the good of the many outweighs the good of the few, holds for damaged cells as well as for Vulcans. An early altruistic death will kill an invading virus and protect the rest of the organism. But in the battle for survival, viruses, for example, have evolved countermeasures. They produce anti-apoptotic factors. The balance of power between microbe and host depends on whether the virus can stop the cell from committing suicide before it has time to replicate.

Programmed cell death acts as a fail-safe mechanism. Killing a mutant cell could nip cancer in the bud. Similarly, just one cell in which the suicide pathway has gone awry could actually be enough to cause cancer and, potentially, death of the whole organism. More often, tumors are caused by genes that normally promote cell growth being active when they shouldn't. So-called oncogenes come in many forms and each cancer has its own hallmark pattern. Strong control over cell division is crucial, because uncontrolled growth consumes huge amounts of energy. Individual cells are cheap, but cancer is costly. In economic terms, it wastefully builds unwanted, unnecessary cells while the rest of the organism starves.

Death is good for you. It is important, not just in warding off infection and nipping cancerous cells in the bud, but at many stages throughout life. It is the carefully controlled mechanism by which we grow, maintain our form and replenish our tissues. The death of cells underpins the remodeling of organs during embryonic development. Repetitive death and replacement provide a healthy turnover of cells in many organs, including the skin, gut and, notably, the blood. (And, curiously, in spite of all this shedding of dead cells and layering of new ones, we still remain by and large the same—with the possible addition of some extra wrinkles or a few grey hairs.)

To try to form a fetus without programmed cell death would be like a carpenter trying to make a chair without a saw and sandpaper. Cells must die to allow new features to evolve. For example, when a hand develops from the embryo's limb bud the fingers grow but the web of skin between them is programmed to die back. Without apoptosis, we would have skin as thick as an elephant's. The cells lining our mouths die and are replaced every few hours. Without apoptosis, oral infection and halitosis would be the odors of the day. If we didn't lose neurons during development we would have larger but inferior brains.

Woody Allen was right when he said that he could contemplate his own death and still carry a tune. When it comes to apoptosis there is plenty to be cheerful about. 'Death becomes us,' I wrote in my *Independent* article, accentuating all the good things: the way it carefully crafts embryonic development, how it makes us what we are. I also wrote about the other all-important characteristic feature of cell death: its irreversibility. Once the process has been activated, when the apoptotic cascade has clearly been set in motion, it is absolutely impossible to turn back.

WMCK3244@AOL.com   Sent: Mon 12/21/1998 17.38

Dear Ruthie
We are on Delta flight DL012 arriving Gatwick North terminal
at 6.55 a.m. on Dec 29th from Atlanta.
    See you soon
    Love Mum.

The same slowly rotating crowd have hemmed me in for what
seems like ages. Then, despite the random pushing and shoving as
people make their way through the crowd, I feel a light tap on my
shoulder. It stands out from the frequent accidental brushes and
bumps. It's a deliberate touch. It's Billy. But it's not my father's firm
familiar hand, the hand that I know. It is soft and gentle, like a
child's. At another time, my father might have seized the opportunity for a bit of spontaneous tomfoolery. Not this time. No words
accompany the frail gesture, but when I turn to see him standing
behind me the vision is better than I could possibly have imagined.
Better than humor, better than history, better than my twenty-first
birthday.

Tall, handsome and neatly dressed in his navy blazer with its
shiny brass buttons, his thinning hair combed smoothly to one side,
he stands there like an all-conquering rear-admiral returning trium-
phant from a foreign posting. He grins. No wheelchair required, the
smile says. He is still walking under his own steam, still carrying his
own navy blue traveling bag.

We head for the exit where my brother Andrew is waiting outside in the car. After a few steps my father's illusionary well-being
evaporates like dry ice when he slips his hand into the crook of my
arm. It is such a little act. Certainly, we have walked along arm in
arm many times, but it was always my arm in his. On rare occasions
when we went shopping together we would walk like this down the
street, but he never held onto me for support before. He might have

pulled his arm in, keeping my hand closer to his body for my own security, but he was always the ring and I was the hook. Even two short months ago when we went swimming in Sarasota it was like that. Now it is the other way around and I can barely feel the weight of it, so light is his touch.

'How was the flight, Dad?' asks Andrew as we adjust his pillows in the car.

'No problem at all, actually. I thought it would be really uncomfortable but it was much better than I was expecting.'

'Your father slept the whole way back. Didn't even wake to eat!' adds my mother.

'Did you get in touch with Marcus?'

'Yes—well, with his secretary. He is on holiday and not back until the third of January, so I made an appointment for 9 a.m. on the fourth.'

My father exhales light and long and makes no further conversation. Weariness or disappointment? Whichever, he is soon snoozing, giving my mother a chance to fill in the gaps.

'He was determined not to arrive in a wheelchair, you know, Ruth. We came through customs in one of those electric carts and right at the door he insisted on getting out and walking.'

I look over at the seat in front of me in admiration. His lips are barely parted. His breaths are barely breaths. His skin is old and pallid yet the wrinkles in his cheeks are new and deep. He must have calculated the limit of his energy and spent the lot on the appearance of well-being to us. How very like him to save up for one big impression. 'Do it once and do it properly' could have been his motto.

My mother twiddles the corner of her tissue into a knot. 'One day he just said: "Katie, I think it's time to go home."' A fragment breaks off and she rolls it around in her fingertips. 'I couldn't persuade him to leave sooner. He wanted to see Stuart, Janet and the kids over Christmas and you know what Billy's like once his mind is made up.'

As the departure date approached, she told us, his mood fluctuated. He walked slowly from room to room, spent hours gazing at the flowers on the deck and the birds at the feeder. When feeling energetic enough, he made plans: he canceled their Internet subscription, arranged a taxi to Sarasota airport, a wheelchair for the plane transfer at Atlanta and again on arrival at Gatwick, so that he wouldn't have to stand in line for hours as he did on the way over. He fed the orchids and then, late one afternoon, when everything was sorted out, he asked to be driven down to Nokomis beach for a last visit before they came home.

She collects the pieces of tissue into a ball and puts the shredded mass back into her handbag.

We learned about death from studying worms. Not the fat, juicy variety you find squirming around after the rain or on the end of a fishing rod, but a millimeter-long roundworm called *Caenorhabditis elegans*. It lives an anonymous existence in bacteria-rich, rotting vegetation; it has done so for millennia and probably would have continued quietly in the same way had the geneticist Sydney Brenner not been so fascinated by its life cycle. His early work led John Sulston of the United Kingdom's Sanger Center and Robert Waterston of the Washington University School of Medicine in St. Louis to decipher the worm's genome in 1998, making it the most celebrated creature in biology. It was the first creature to have its DNA completely mapped and the contents of its genome provide a complete set of instructions—a compendium of everything needed to make an animal. *C. elegans* is made from 19,099 genes—almost as many as humans. The pattern of cell division in the worm, from fertilization to death is now understood, earning John Sulston the Nobel prize in 2002. Every synapse has now been tracked, every gene studied, and along the way the principle of apoptosis was discovered.

Normally *C. elegans* takes three days to reach maturity and an adult worm is made of 959 cells. Without one of the genes that is necessary for apoptosis a worm can reach 1,090 cells. You might imagine that removing a gene for cell death would make an animal live longer. Not so. When apoptotic genes are lost or mutated, the roundworms do not live a day longer than their normal counterparts. Instead, they have exactly 131 extra cells. By disrupting one gene at a time and searching for animals in which the death of those 131 extra cells was interrupted, MIT's Dr. Robert Horowitz was able to find the genes necessary for apoptosis, an accomplishment that won him the Nobel Prize for medicine in 2002. We know now that there are at least ten genes involved in apoptosis (ced-1 to ced-10), annotated ced because the worms are abnormal in *Cell Death*.

These crucial genes rarely make an appearance under normal conditions when a cell is healthily going about its usual business. They become active in times of trouble or, at least, when circumstances dramatically change. And the more important ones (ced-3, ced-4 and ced-9, if you really want to know) have specialized functions; they are like the property developer, the planning officer and site foreman of a redevelopment site. The property developer puts the plans together and is unable to proceed until approval is confirmed by the planning officer. Once the go-ahead is received, the site foreman helps his men find the easiest way to take the place apart. Ced-3 instructs ced-4 and then ced-9. There can be no other sequence—it forms a cascade of actions that control the whole pathway of deconstruction.

My father's first visitor after arriving home is the local general practitioner (GP), a chubby man in his fifties with a cheery round face, slightly pink complexion and receding curly grey hair. Chris Pinchen is less of a doctor and more of a friend as a consequence of his decade-long association with Molecular Products. He is the very incarnation of W. H. Auden's ideal doctor.

> . . . *partridge-plump,*
> *Short in the leg and broad in the rump,*

An endomorph with gentle hands,
Who'll never make absurd demands
That I abandon all my vices,
Nor pull a long face in a crisis,
But with a twinkle in his eye
Will tell me that I have to die.

Dr. Pinchen accepts Mum's offer of a cup of tea and, after examining his patient, cradles the mug in both hands and reassures her that they have absolutely done the right thing in returning home. He suspects something untoward, and the next step will be to consult with Dr. Marcus, arrange a visit to Addenbrooke's, and thereafter he will orchestrate whatever plans are necessary.

While my father waits for the appointment with Dr. Marcus he is 'at home' in the most Victorian sense of the word. He is available for all comers, and when one golfing chum asks why he returned early he replies off-handedly, 'Well, I wasn't making the progress I had hoped for.' He says it in a way that implies he can only run ten miles instead of twenty or manage a mere dozen press-ups instead of the usual fifty and he talks of going back to see his consultant as if it were a routine checkup. His nonchalance is as convincing as a *Reader's Digest* letter proclaiming that you've won £25,000. Only the most gullible would readily believe it.

There is no public discussion of my father's condition, nor has there been since he returned. Together we speak in generalities, of tiredness and lack of progress; words that end in -itis or -oma are not mentioned in his company. In an effort to open up some dialogue on the matter I ask him what he thinks is wrong, but he replies with irritation, as if he would rather be lying on Nokomis beach, watching the pelicans, than having this conversation. From a distant place he says, 'Oh, I don't know, Ruthie. Perhaps I've got hepatitis. I feel a bit liverish.'

I nod in agreement although I have no idea what 'liverish' might mean to him. Is it possible to be kidneyish, brainish, lungish? On reflection, I don't really want it to develop into a conversation in which I might have to admit that the unspoken secret in the kitchen is liver cancer. Chris Pinchen found multiple nodules while carefully feeling his patient's abdomen, but nobody will say the words out loud, not until he has had the scan. He doesn't have cancer until the shadow appears on the film. He doesn't have cancer until there are data, measurements and facts.

'A bowl of soup would be nice.'

We make enough to feed the starving masses of the Sudan. Each batch is of the Scotch broth variety, like my grandmothers used to make with 'neep and lentils, beans or barley. He starts out by eating a big bowlful in the kitchen with a fat slice of bread and enjoys some conversation with the rest of the family. But over the next few days the portion becomes smaller and he eats on his own, resting the bowl on a small table alongside the sofa, in a way that he would have disapproved of had we offspring done it. He leaves the bread untouched. The day before the scan he sups the liquid only, rejecting the vegetables. He does so without lifting his chin from his hand, opening his mouth by raising his head, rather than lowering his jaw.

Tuesday, January 14, and the trip to Clinic 2 at Addenbrooke's is classic farce. The moment my father steps out of the house and into the sunlight, it hits me. His skin is bright yellow. Jaundice is characteristic of liver damage caused when bilirubin, a component of old red blood cells, can't be broken down by the liver—as it usually is— and the yellow pigment builds up in the body. His skin's glowing translucence makes me think of the film of oil on the top of a take-out curry. He moves slower than an eighty-year-old performing T'ai Chi but he eventually climbs into Andrew's car. I prop him up with three or four pillows, further padded by a duvet, and watch the tension in his face record every bump of the twenty-five-minute drive to the hospital. By the time we arrive, he has the strength and resilience of a well-washed duster. Andrew and I manhandle him into a decrepit wheelchair and push it, crab-like, to the nurse at the reception desk.

'Now, Mr. McKernan, you need to drink this and wait over there for half an hour,' she says, handing him the two cups of liquid radionuclide that will illuminate his scan and seal his fate. There is as much chance of him sitting upright in a wheelchair for thirty minutes as there is of him winning Olympic gold for the pole vault. I plead for him to be given a bed like all the others in the room. They are admitted patients lying down comfortably and he sits in the middle of them, a buttercup in a clump of daisies. The nurse on duty looks at us keenly, makes a couple of calls and allows us to use a temporary hospital trolley-bed in an adjoining room. We take off his new American sneakers to spare him the effort of lifting their weight onto the bed. We try to keep him occupied, talk of inconsequential things, but every conversation ends up asking how he is feeling and we hush so that he can concentrate on keeping the foul liquid down, for if he vomits, tempting as it is, he will have to start from the beginning again.

After the scan a portly, middle-aged man in a white coat introduces himself as the radiologist. His name passes me by because my attention is fixed on the films in his left hand. I expect him to put them on a light-box and highlight the features of interest. He says with some self-satisfaction that they have managed to get images of high quality.

Then show me, shouts a voice in my head.

He doesn't. Instead he chooses words rather than pictures, keeping the films tight against his chest. 'There is extensive abnormal tissue in the liver, lungs and abdomen,' are the ones I commit to memory. There are others: promises to discuss all options with Dr. Marcus, confirmation that Dr. Pinchen is his GP, surprise that my father is an outpatient. There will be discussions between Dr. Pinchen and the radiologist—the property developer and the planning officer. There will be a path ahead.

We have heard enough. We parcel my golden father back onto his feather bed and take him home.

His GP friend returns the next day to confirm that the cancer, which may have started in his pancreas, has now spread to many other organs. Without losing the kind twinkle in his eye the gentle-handed endomorph gracefully closes the door.

We might call ced-3 the molecule of no return but what does it actually do? In 1993 several groups of scientists, among them Don Nicholson, a colleague of mine from Montreal, found that the *C. elegans* gene, ced-3, is the template for making an enzyme in the worm with remarkable similarity to one already known in mammals. The enzyme's job is to cut up proteins. Closer comparison of the worm genome with the emerging human genome revealed that there are about a dozen similar enzymes, forming a family called caspases.

All cells contain caspases and they are normally present in an inactive form, to prevent unregulated self-destruction. These molecular scissors are responsible for trimming or sequentially chopping up the protein components of cells. Some caspases activate others, like one pair of scissors cutting away the packaging around another. Together, they perform the cell's careful deconstruction. It is a neat, well-controlled job. There is an order in which things happen and taking a cell apart requires as careful a sequence of events as cell division. When cells divide a whole cascade of genes organize chromosomes to be duplicated, separated on a spindle and segregated to different ends of the cell. When cells collapse a different cascade of genes comes into play, expertly guiding the cell through multiple checkpoints until the new set of rules directs only its demise.

Dr. Pinchen is quite clear with my mother and me about the extent of my father's cancer and the advice 'Nothing more can be done' leaves little room for ambiguity. But we don't know exactly what he said to his friend. Has Billy worked it out? His own father died of cancer and he watched him get weaker and thinner. Billy is a clever man, but you don't need a Ph.D. to assimilate the evidence:

his genetic history, the weight loss, his current infirmities, the way Chris Pinchen examined him. Does he know? Does he think there is still hope? Does he realize that the bottles of relaxing Radox bath foam, multiplying on the shelf behind the whirlpool bath because he once commented that they relieved the pain, will never be emptied?

To acknowledge what we both know is the hardest moment of all—harder than any hospital visit, harder than sitting by his unconscious body or watching his frail attempts to walk. I don't want to cry—and certainly not in front of him, not now. I quietly go upstairs to see him. I ought to have been well prepared for this moment. It's not like I didn't know it was coming. I might even have felt grateful that my three wishes for him had been fulfilled. After all, he did get to see Scotland and Florida again and to celebrate Molecular Products' twenty-five-year anniversary. I don't think about what I will say to him, only about how to maintain my composure. The right words will surely come provided that I can keep the tears from dissolving whatever it is that is holding me together. Up the last few steps. The world is no different than it was yesterday; he hasn't suddenly got worse. It is twenty-four hours closer to the end. But that is true for all of us. Every day is one step closer to death.

Today is no different from any other day; as Samuel Johnson said, 'It matters not how a man dies, but how he lives. The act of dying is not of importance, it lasts so short a time.' I try to convince myself that nothing has changed. Except, of course, the difference is that all hope is gone.

I clench my jaws together. I purse my lips. *I can do this, I can do this.* I open the door and go in. Still, the tears come.

'Oh, Ruthie, please don't,' he says.

The agony in his face holds everything. I have been so busy thinking how hard this will be for me that I haven't thought about my father's struggle. One request is enough and I don't know how it happens. Perhaps I am so accustomed to carrying out his wishes that the tears automatically slip back into their ducts. I stop. Simple as that! I take a few breaths and join him in the displacement activity of rearranging pillows.

There is no poignant conversation between daughter and dying father, clutched in tearful embrace. Hollywood melodrama is not for us. If there were a time to reveal my innermost feelings, this would be it. Had my father wise words of momentous import for me? Now would be the time for him to share them. Do I have any final confessions? Would he really care that it was me who opened the packet of chocolate digestives, without asking, thirty years ago? And as the moment passes, I feel no loss. Those unsaid things don't need saying because, as any behavioral psychologist—particularly my husband— will tell you, it's not what you say, it's what you do, that matters. And I have done all I can possibly do for him and it is nowhere close to what he has done for me.

At times like this people revert to type. Organizers busy themselves in planning. Worriers worry. Some families unite, while frailer groups fall apart. My family becomes slime mold. This single-celled creature, properly called *Dyctostelium discoideum*, usually lives life as a multitude of separate cells, each independent of the others. But in times of stress, individual cells coalesce to become a more complex, multi-cellular organism in which the original cells group together and take on specialized functions. So Mum becomes social secretary; Ian turns into treasurer; Stuart is sucked back in from Minnesota and covers a plethora of roles as an active committee member during the day while the rest of us try to go back to work. Andrew collects medicines and runs any errand my father requests, fitting it all around his job at Molecular Products, which seems to be going OK. As for myself, the role of chief clinical interrogator and medical monitor is now redundant so I become daughter again; not scientist, not quasi-quack, I'm maybe something between rookie nurse and free entertainment.

Looking after a dying man takes training. Since my father's illness, I have opened the *Oxford Textbook of Medicine* many times, but

I have lacked the courage to read the chapter entitled 'Terminal care' by M. L. Baines. Now I consult the text like the Bible.

There is a general understanding that terminal care refers to the management of patients in whom the advent of death is felt to be certain and not too far off and for whom medical effort has turned away from therapy and become concentrated on the relief of symptoms and the support of both parent and family.

'Comfortable' becomes a most overused word. I adopt it like a stray cat, showing it to visitors for empathy. 'He's comfortable at the moment.' 'He had quite a comfortable night.' 'Yes, now he has stronger painkillers he's much more comfortable.' At each stage of his deterioration we justify whatever intervention is necessary as an offering to the god of comfort. Dr. Baines sets out very clearly how terminal care at home is preferable to being in hospital: fewer bed sores, less anorexia, better sleep. And, not insignificantly, at least at home you can get a decent cup of tea when you want one.

Symptomatic relief and pain management are absolutely necessary but such nursing skills and knowledge, unlike infantile adoration and teenage truculence, don't come hardwired in a daughter's brain. What should be simple tasks prove to be beyond my expertise, beyond the whole family's experience. We can't change the bedding fast enough. We ineptly fumble around changing a sweat-laden sheet for a fresh one while leaving my father leaning on a bedside chair, trembling with the effort of standing, and clutching at the cord around his trousers to keep them from falling down. The practical polyester pajamas that fitted him just nine months ago now look like they've been borrowed from Pavarotti. What do we do when he can't drink from a mug? We need help. Professional help! We need someone who knows how to manage the inevitable painlessly.

Nora Roberts, the district nurse, is our site foreman. If my father could have had his pick of attendants, she would undoubtedly have been his choice. This lively Irish woman in her late thirties with neat layered hair and creamy pink lipstick makes no assumptions. She allows all the decisions to be my father's, talking to him as if he were not an invalid. Control is diplomatically restored, giving him the sense that he is advising her on the best way to do her job.

On her first visit she starts off with: 'What would you like me to do?'

'Tell me a joke,' he replies.

Without hesitating, changing the tone of her voice or questioning his reply, she parries back: 'What would I be if I weren't Irish?' She gives my father a minute or two to mull over whether she has taken him seriously. 'Ashamed,' she says, and she laughs one short sharp laugh, setting the tone for the rest of her visits.

Most days start with a joke. Nora is not intimidated by my father, although, in keeping with his lifelong habits, he treats her almost as an employee. He demands conversation on his terms. He wants answers, but only to the questions he asks. It is nursing within limits. He wants to know what the medicines are and what they are for. One by one, she goes through the containers lined up on the melamine tray beside his bed. There is Valium if he struggles to sleep; a morphine solution, Oromorph, to relieve the pain, which is getting worse; and glucose syrup to loosen the bowels, counteracting his constipation, a side effect of the morphine. When he can no longer swallow the Oromorph, Nora puts in a syringe driver, which injects the opiate directly into the top of his arm every two minutes or so, but he rarely complains. Occasionally he says his shoulder is uncomfortable and we buy a heat pad which he wants on and off with uncharacteristic indecision. When he can't sleep from the pressure on his bones, Nora brings an air mattress and some cream for his thinning skin. When her required tasks are done they share a cup of tea and a biscuit and she sits close to him in the way that his PA Carole used to when reviewing her upcoming assignments.

All the same, Nora knows that he will share no intimacy with her. Washing, and anything else of a personal nature, is still strictly my mother's domain. Others might let their nurse handle everything, but my father is not programmed for such behavior. He will not collapse in a frenzy of fear or be forced, through sickness, to abdicate responsibility. Necrosis is for those who can't help it. While Billy has the mental capacity he will not relinquish control.

Apoptotic cells all die in the same way. Before they can fragment they must extricate themselves from the surrounding tissue. Any interactions with distant cells are minimized and their fluid content, or cytoplasm, condenses. Apoptotic cells stop integrating or contributing to the processes around them. They carefully withdraw from the complex communications they once had, shrink and round up in preparation for their own demise.

With little railing at the fading light, my father deliberately begins to dismantle the life that he has taken so much pleasure in building. He says he would have endured a return to Addenbrooke's for further tests or even surgery if there were a small chance that the medical establishment could do something for him. Before I have time to think of an appropriate comment—for I know that checkpoint passed some days ago—he interrupts my discomfort, saying, 'But they'll get all the information they need at autopsy.'

'Oh, Dad!' I put my hands lightly on his shoulders and move my face closer to his. He reaches up and tries to hug me. The once crushing bear-hug has withered in force, his arms curling round me as a long blade of grass hugs a tree.

He draws up a list of the people he wants to see and on good days manages two or three visitors. His burly angler friend Keith Johnson arrives from Aberdeen bringing with him a side of smoked Scottish salmon. His arrival blows away the thick dusting of sadness that has settled in the house and for a moment there is laughter as the two of them reminisce over past fishing trips in Perthshire. His grandchildren visit with hand-drawn pictures that are carefully prompted to say 'We love you, Granddad' rather than 'Get well soon' and the youngsters innocently fail to notice how thin he has become, or the strange yellow of his complexion.

Colleagues from Molecular Products troop up the stairs to order, but not before being thoroughly briefed on his current condition and appearance. My father's long-time PA is greeted with 'At last— Carole' (which he pronounces as a three-syllable word, Ca-ro-lee, in the way he often did when teasing her). 'I've been trying to get you into my bedroom for fourteen years and now that I've finally managed it, I'm in no fit state to do anything about it.'

John Addison uses his allocated time to update his former boss on what's been happening at the company. Initially he avoids the subject, bemused that such a sick man would be interested, but my father insists.

'I'm fed up hearing how ill I am, John. Tell me what's going on.'

So the conversation with his mentor (I learned a lot from your dad, John once told me) is much the same as on any working day. They discuss the current state of negotiations for the new contract with the American navy. They mull over the problems in production and they try to anticipate when the next big order is likely to come in. It's as close to independent reassurance of the company's longevity as my father is likely to get. He could have had the same discussion with Jim Beard, the accountant, but Jim has less to say and stays a shorter time, leaving with the words: 'If there's anything my team can do for you, don't hesitate to call.'

'Thanks, Jim. See if they can find a cure for cancer—it's pretty urgent,' the devil on my shoulder silently retorts. But my father says nothing of the sort, wouldn't even think of it. Instead he smiles, thanks Jim for coming, asks me to turn on the heating pad again and drifts off to sleep.

As my father becomes less predictable, visitors are reduced to family only. To keep him entertained we do the *Telegraph* cross-word. One clue gives us particular trouble. Five down:

bird droppings (four letters). There is a pause. Then my father's face breaks into that mischievous smile, once frequent and easily provoked.

'Shit?' he says.

'I don't think that can be right, Dad—not in the *Telegraph*.'

We complete a few more clues and return to five down: bird droppings (four letters).

'It's eggs,' he says, having known it all the time, and laughs.

There are other, less convivial times when you wouldn't want to send anyone up to see him, when my mother and I hesitate to go ourselves, knowing that we won't be able to avoid walking on eggshells. We want to keep him comfortable, to sit at his bedside and pander to his needs.

'You are like the old ladies of the French Revolution. Don't you have some knitting to do?' he says, without a hint of humor in his voice.

Having withdrawn connections from their neighbors, apoptotic cells begin to disintegrate. There is elegance in apoptosis. Under the microscope, cells go through their well-ordered moves with the timeless choreography of a Busby Berkeley dance sequence. Unlike the process of necrosis the cell's membrane doesn't swell and burst; rather, it shrinks, forming blebs, small spheres of material on the surface of the cell, which become detached and float away. The nucleus shrinks into a dense body in the center of the cell and then breaks into small fragments that get distributed, finally ending up in the blebs, along with other cell components, most of which are still functional. New signals are put out on the surface of the disintegrating cell. 'Take me, I'm yours' is the message to its neighbors. Healthy young neighboring cells engulf and absorb the fragments in a process known as phagocytosis. There is no spillage, no inflammation, and the troubled cell quietly disappears into the surrounding tissue.

Other new enzymes, called nucleases, come into play. Nucleases are the enzymes responsible for chopping up DNA. Once the cell membrane disintegrates, nucleases can get access to the nucleus and break it down too. Fragmentation of chromosomes can be seen through a microscope as the emergence of characteristic dots where one solid dark sphere of a nucleus used to be. Nucleases are rarely active during a cell's normal lifetime, but they are critical to the execution of apoptosis, to pass the cell's contents to others who can make most use of it.

Billy's three special-edition bottles of malt whisky are lined up on the table beside his bed. As instructed, Mum has brought them from the study where they had been gathering dust, like old maids, waiting for an occasion important enough to be opened. The porcelain flagons with their pastel hand-painted scenes of St. Andrews are dusted, cleaned and given to a few select golfing buddies. His gold and silver chess set, a souvenir from Mexico, is put aside for my older brother Stuart, the only one of us who was ever any good at the game. The silver Georgian set of decanter and glasses that Molecular Products gave him as a retirement present remains in its box. They have never been used. I guess they'll be for Ian and one day they'll pass, it's to be hoped more smoothly, to whoever takes over the company after him.

Billy examines the details of his life insurance and summons his lawyer to establish that my mother's inheritance is adequate, which it is. He checks that the tax returns are up to date, which they are. The deeds of the family home are recovered from the bank and all important documents are neatly organized into a tidy little folder along with his share certificates, the children's trust certificates and his credit card statements, fully paid up. He sends Ian out to buy a present for Mum's birthday, which is on January 29, just two weeks away. He asks me to call Delta Airlines to enquire whether his air miles can be transferred to my mother's account. I procrastinate. What's the point in doing it straight away? He can't die until everything is sorted out, and if there's still unfinished business he can't die.

The process is irretrievable, I know that. Dr. Pinchen, the nameless radiographer and Nora Roberts are property developer, planning officer and site foreman. They ensure an apoptotic death just as surely as ced-3, ced-4 and ced-9. His life is being carefully fragmented and disassembled and it is my job not to complain, but to contribute, to support my parents in seeing this through with the minimum of fuss; a predestined, inevitable pact.

It has to be this way, for death is good for you, or, if not good, then necessary. I know that. Scientists have believed it for centuries. Long before we understood anything of the sophistication of apoptosis, in fact two hundred years before Charles Darwin wrote *The Origin of Species*, the French scientist and philosopher Blaise Pascal likened the human race to a man who never dies. We need to turn over continuously, evolving slowly as we go to ensure the survival of our species. The death of any one individual is no more than the apoptosis of mankind. I understand that sensible scientific principle in the abstract. Birth and death ensure the survival of a species; any scientist acknowledges the need to balance that equation. But that is science and this is real life. In a parallel, emotional universe lives this daughter with her fears. Grief is just another form of pain—and all the theory in the world doesn't make it any easier to bear.

11

Time

I am forced to look at time
In another way. Not
As so many grains of sand
Flowing from glass belly, but how,
Through the persistent gnawing
Of years, I've weathered
As I watch them grow. And
How at last, as I let go
And slip behind them, I ease
My bones into the universe.
From 'Letting Go' by Angela Greene

MY FATHER IS AWAKE.

'What day is it, what time?'

He moves his head marginally to look at the electronic clock beside his bed. He seems puzzled by the display and dozes off again.

Awake again a few moments later, he has the same question on his lips.

His conversation has become staccato. Coherence started to fade a few days ago when he occasionally replied inappropriately when we were talking, as though he hadn't properly heard what I said. And when it was his turn to speak, a few words were missing from his sentences. To hide the gaps he developed his own brand of frugal speech, communicating only in short phrases, having given away most of his vocabulary.

'What's the time?'

The question is unnecessary because his bedside table has more ways of measuring time than a sick man would need or want. Alex, my father's long-time legal adviser, even gave him a calendar last week. He'd been to an exhibition at the London Design Center and thought his friend might like the souvenir. It is indeed a work of art with origami-style inclusions that burst out each month and under other circumstances, yes, he would have been most appreciative. It sits beside the bed, the paper cotton reel and needle standing out over January/February. For what reason would a man who had his last bath five days ago, his last shower on Saturday and made his last trip to the bathroom yesterday need a calendar? And it's not the numbers with imaginary lines through them that bother me, it's those that will never be defaced.

A clock radio sits alongside. I read the digital display out loud. 'Four thirty in the morning, Dad.'

His brows furrow as if this isn't an answer to his question, as if time has lost its meaning. It is no longer the even linear progression introduced in 1582 under Pope Gregory XIII. The Gregorian calendar means nothing anymore. To the rest of us, time marches forward, seeming to accelerate as we age. It is Shakespeare's 'eater of youth' that 'nursest all and murder'st all that are'.

Imagine if we could wind time back. Does it really need a constantly forward trajectory? It hasn't always been considered a straight line. For example, the Mayans of Central America saw time as a series of cycles, represented by recurring spells of duty for each of their gods. Theirs was a civilization obsessed by time, each day divine. Their altars and monuments honored time, not people or rulers. Their method of measuring time was more accurate than ours too. (In every 10,000 days the Gregorian calendar is three days too long, whereas the Mayan calendar falls only two days too short.) Each time cycle was expected to repeat itself in 260 years, when the same gods would march together again and events would follow a similar pattern.

Where would my family be next time round? Who will be calling out: 'Ruth, where's Mum? What's the time?'

'She's asleep. Should I get her?'

He stares past me, then closes his eyes. Asleep again, as he should be at this time of night.

We take for granted that we wake in the morning and go to sleep in the evening; that our bodies are neatly aligned to the time of day. When matching physiology or behavior with time, we are no different from the pet hamster I had as a child. You could have set your watch for 10 p.m. when Hopalong (well, the hamster had only three feet) started trundling around in his exercise wheel. All earthlings are governed by the planet's variable periods of darkness and light. All animals, whether they are active during the day or night, use the same biological processes—the same genes—to synchronize their bodies to the external world. Insects and even single-celled organisms that live for less than twenty-four hours have some sense of the time of day. My father, no longer.

The circadian system (from the Latin *circa diem*, meaning 'around the day') manages our cycle of sleep and wakefulness in response to external signals. Light, physical activity and even social interactions help us keep a well-ordered rhythm. In a healthy person, the circadian cycle takes twenty-five hours to complete, not twenty-four, as you might have thought. Volunteers, kept in darkened rooms, found that their natural sleep–wake cycle extends the day by an hour or so.

There are few inventions of man that nature didn't find first by evolution. W. H. Schott built a clock in 1921 which was upheld as a masterpiece of design, a revolution in timekeeping. It worked on the principle of having two pendulums, a master and a slave. The master pendulum drove the timepiece and the slave pendulum maintained the master's motion. The clock's natural tendency would have been to run slow were it not for the slave pendulum's constant priming. The principle is not so different from the human time clock. The brain's circadian center (the suprachiasmatic nucleus, or SCN) acts as a master pendulum and it would keep the sleep–wake cycle at about twenty-five hours were it not constantly adjusted by light—the equivalent of the 'slave pendulum'—to the periods of day and night in the environment. This dual system would keep the body regular while allowing for adjustments in the length of the day, from winter to summer, as Earth revolves around the Sun.

Nowadays, we are primed by alarm clocks and lightbulbs rather than by the length of the natural day. But before we had artificial light and accurate, measurable units of time, days were truly longer in the summer and shorter in the winter. The timeframe we accommodated was measured in days and nights, lunar months and seasonal years, not hours, minutes and seconds.

My father no longer recognizes any of these. He barely knows day from night, and time, for him, is not measured in the same units as for the rest of us. The numbers on his clock radio are no more than angular patterns of liquid crystal, and time, that most precious of commodities, a thing he rationed and monitored with care, is dissolving before his eyes. Sleep and wakefulness roam free, unfettered by biological constraint, like the rhythmless cycle of a newborn baby.

How does the suprachiasmatic nucleus control the body's timing? Actually, I should say nucl*ei*—plural—as there are two of them. Each comprises a cluster of about 20,000 neurons, collectively no larger than the winder on my father's watch, and they're sited one on each side of the brain, way under the cortex, about as deep within it as you can get. The amazing thing about the SCN is that it generates its own phasic pattern of electrical activity, firing synchronously eight to ten times a second during the day and two to four times a second at night, even if the cells are removed and put in a laboratory dish. Few other cells independently maintain their own rhythm (the obvious exception being cardiac cells, which similarly continue to beat when isolated). Like all else in biology the SCN's function is defined by genes. 'Clock genes' have been discovered that form a self-sustaining timepiece in species from fruit flies to man. These genes build up the machinery that turns on wakefulness, activity and sleep, and all else that goes with it. The details of how the genes maintain a twenty-four-hour cycle is fascinating. Unlike most other body parts, which are built to last, clock genes give out temporary instructions to produce a short-lived product, like an instruction manual written in fast-fading ink, or a clockwork assembly built of ice. The cogs melt away each day, only to be remade when new instructions appear twenty-four hours later.

Many cells have circadian oscillations and all are linked to the SCN, which sets the pace. Without a functional suprachiasmatic nucleus an animal wouldn't be able to control the oscillations in its neurons, nor indeed in any other cells. The SCN is the André Previn of the body, conducting the symphonic activity of different organs. Without a functional SCN, rhythms across the whole body desynchronize and collapse like a discordant, leaderless orchestra.

In the days when my father was building up Molecular Products he was a human dynamo with only two speeds of operation: right now and very, very soon. Some people are 'morning larks', like Billy, up early and straight onto the golf course, or into the traffic line for the Blackwall Tunnel, while others can barely lift their heads off the pillow by the crack of ten without a cup of strong coffee. Night owls compensate by being alert and fully functional long after the larks have had their cocoa. Disrupted clock genes can cause unusual circadian rhythms. Research has uncovered a family in Utah who have a mutant clock gene. It melts too early, making them the earliest morning larks on the planet, waking between 2.30 and 4.30 in the morning with a natural bedtime of 7.30 p.m. Two thirty or seven thirty, it is all incomprehensible now to Billy. His watch and clock radio mark only the sporadic nature of sleep.

My father scrabbles about in some dark space, pulling out things that he half recognizes. There is something of great importance in there somewhere. He just can't seem to remember where he put it. 'What's the time?' he asks repeatedly and with increasing agitation. He tries different paths but his brain consistently denies access. 'The time!' he shouts, if shout is an appropriate word to describe the widening of his mouth with no increase in volume. 'What's the bloody time?'

Why the obsession with time? My younger, fitter brain should know what he's looking for. I should be able to work out whatever it is that's troubling him. My father is gripped by an imperative and urgent need for an answer to his question, and that answer isn't four thirty or noon or midnight. If the obvious answer isn't the right answer, then maybe the question he is asking isn't the right question. With a fractional vocabulary, maybe he can't find the question. What is it that he means to say?

His bony fingers feel around the bed. They slip between the air mattress and the proper mattress below but find nothing. Tugging at his pillow doesn't seem to help and he shakes his head in slow motion, dropping his arm back to his side.

'Dad, do you need something? Can I get you anything?'

'What? Not now. Don't know.'

Sleep. More sleep. You need more sleep to find the energy to make the words to say what you want to say.

After a good sleep, activation of the HPA axis (the hypothalamic-pituitary-adrenal axis) prepares us for the mental and physical demands of the day. Blood pressure rises and cardiac muscle prepares for action. At dusk our temperatures begin to fall and melatonin is released to facilitate sleep. Hormones that control growth and development, such as growth hormone and prolactin, which build bones and muscles and encourage reproductive maturity, are also released at night. Mick Hastings, from the Laboratory of Molecular Biology in Cambridge, has studied which genes are under circadian control. Fortunately, the recent explosion in molecular biology has provided the laboratory tools to help him. Gene chips (also called gene arrays) can detect every active gene out of the tens of thousands in an animal's genome. A pregnancy-testing kit can detect just one hormone, whereas a gene chip—still no bigger than an after-dinner mint—can detect 25,000 active genes from just one drop of extracted blood or tissue.

By taking cells from different organs of the body at different times of the day (and, unfortunately for the researchers carrying out the work, the night too), Dr. Hastings and other scientists have found that 5 to 10 percent of all genes in the liver are under circadian control. Genes that make the enzymes responsible for metabolizing food are more active during the day, ensuring a ready energy supply, while those that make fat and store energy work harder while we sleep. During daylight hours, genes that matter for interaction with the world dominate; those that activate brain processes, feeding, reproduction and physical strength have the upper hand. At night, growth, repair and memory consolidation dominate. Dr. Hastings calculates that across the entire body most of our genes are subject to circadian control somewhere. The circadian system keeps the lights on and the fires burning during the day and closes down the factory for maintenance and stock-taking at night.

In the kitchen, hands clasped round yet more cups of tea, we scrutinize Billy's condition and behavior.

'What's he looking for? Why does he always want to know the time?'

'All he's wanted to talk to me about is Mum's birthday,' says Ian.

'Still four days to go. Do you think he can make it?'

I shrug my shoulders. Truly, I have no idea.

Maintaining an internal clock requires energy. In sickness, energy is at a premium, and when cancer steals whatever it can access each bodily function must justify its survival. In the battle for dwindling resources, circadian processes may well lose out.

Cancer cells, like other cells, are under the baton of the

suprachiasmatic nucleus: initially they do not grow at an even pace; they grow faster at some times of the day. Knowing which genes are active, and when, is beginning to influence medical treatment, which may be more effective when aberrant genes are most vigorous. Furthermore, tumor-suppressor genes respond to DNA damage according to the body's clock, with repair or apoptosis taking place while the patient sleeps.

Loss of circadian control has been used by doctors to gauge survival time in those with advanced cancer that has spread to many organs. Once the body's rhythm of sleep and wakefulness is lost, decline accelerates. The cancerous cells' cycle dominates, stealing the body's energy to set their own unrestrained pace.

For my father, all other things are in their place. The mental list of things to do before he slips into the universe has been efficiently ticked off; all financial and business matters have been resolved. Just one task remains outstanding. With days to go until what has always been the big annual family celebration (possibly surpassed only by Christmas), there seems little possibility that he will survive to see his wish fulfilled. We all know what he has planned. As usual we are his confidants; for he loves nothing more than to surprise his wife on her birthdays, with the rest of the family complicit in the act. It is so very like him, so very Billy.

He would pride himself on his choice of gifts. 'I'm taking your mother on a cruise for her birthday,' he once casually dropped into the conversation. Then there was the Christmas when he bought her emerald earrings and hid them in the pocket of an apron, which he wrapped and put under the tree. He tried to persuade her that the apron was her present, but she knew him better and held out for something more. He kept repeating, with mock irritation, that she had had all she was getting and only several hours later did she discover that the unromantic functional present was the wrapping, not the gift. And the following year, to carry on the theme, he gave her oven gloves and, cruelly, they contained no precious stone. We helped her scour the house for the trinket we knew must exist but the family's exhaustive search turned up nothing. And then she found her ring, disguised as a Christmas bauble, hanging from the tree.

Those events could have been yesterday, last year or a decade ago, but one thing I know is that they are in the past. The human mind is unique among animal species in that somehow, during evolution, we have developed the ability to travel mentally backward and forward in time, blessed as we are with episodic memory. And we have the ability to imagine the future. The average four-year-old can manage the rudiments of mental time-travel, understanding the concepts of yesterday and tomorrow, without which it would be difficult to reason or plan.

Without mental time-travel, without 'last week' or 'next year', our ability to operate in this world is severely blunted. The devastating cost of living only in the present was poignantly demonstrated by the British musician Clive Wearing. Encephalitis left him with no memory of his past and though he understood timing and could still conduct an orchestra he has no concept of past or future. His diary recorded that at 3.20 he had just regained consciousness and the great relief he felt being back in the living world. On the next line down, 3.20 is crossed out and replaced with 3.25, with the same sentiments recorded and so on, repeatedly. He is stuck in the present with no understanding of moving through time.

Perception of time is flexible when we are awake and even more so when we sleep. Dreams that appear long and complex may last only seconds and much of sleep is not associated with any mental perception of time at all. When my father was first admitted to the hospital, each minute spent waiting for the doctors to arrive was stretched by fear. Survivors report experiencing an accident in slow motion, and the belief that your life flashes before you if you jump from a high building may yet have some element of truth. Imminent, intensive fear slows the perception of time, just as heightened pleasure can accelerate it. The present is all we truly experience and

time, like beauty, is in the eye of the beholder. It can pass like the
Tokyo bullet train or it can drag on forever, like overtaking a cyclist,
uphill, in a Fiat 500. Our minds take little account of hours, minutes
and seconds. Perception of time doesn't bow to the quantities measured by any chronometer. It is a malleable, flexible state that
we inhabit where a moment can be a lifetime and a lifetime, just a
moment.

My brothers are sitting by Billy's bed.

Ian holds my father's attention, speaking slowly and deliberately.
'I brought the present with me. I think you should give it to her,' he
says, every word weighty and clear.

'In case you don't feel well enough on Friday,' I nervously add,
just in case he might realize that it isn't January 29 yet.

His empty gaze refocuses and he nods resignedly. 'Go, get your
mum.'

There is no warning that she is walking into another of his little
schemes, no hint that the rest of the family all know something she
does not. Except that we make a space for her right next to him and
the scent of expectation is so heavy in the air that I begin to sweat.
My father slowly takes a little box from under the sheet and gives it
to her, his hand trembling from the effort. I can't hear what he says.
His voice is so weak, he has dispensed with volume as well as
vocabulary.

She lowers her head closer to his and very gently kisses the pursed
yellow lips that present themselves. 'Oh, Billy, thank you.'

She takes the gift and admires the quality gold paper decorated
with crimson tulips. The matching card that is tucked into the ribbon she removes almost in slow motion, to stretch the moment.
Very gently she undoes the bow and takes the paper off reverently, as
though not to hurt it, as if it were part of him. She carefully puts it
aside and opens the black velvet-lined box. She undoes the clasp on
the woven gold strap and takes out an elegant round-faced watch.

'Oh, Billy,' she says again as she strokes his hand and arm. 'Thank you.'

She holds the gold timepiece tightly, not knowing whether to look at it or at him. As she struggles to put it on with one hand, it flips over to reveal the words 'Kate, with all my love, Bill'.

When he sees that she has read it he clenches his teeth and momentarily turns his head away. She won't let him down by crying but does her best to smile, in appreciation of the true magnitude of his gift.

To lighten the mood we explain how Ian brought in a brochure, how Billy picked the one he wanted her to have and then insisted that his chosen words be engraved on the back—so she couldn't flog it after he was gone. We even laughed at Dad's fight to sustain the element of surprise. He had been showing the watch to Andrew and had hidden it back under the bedsheet. When he heard Mum approaching with a cup of tea some hours later he hadn't realized how much time had passed, nor that the watch was safely back in Ian's care. He thought he had lost it and, fearing that she might find it prematurely, he searched around in desperation. His slow-motion actions quickened to a brief frenzy as he leaned out of the bed, pulling bits of paper out of the wastepaper basket in case it had fallen in there. He moved with more urgency and speed in those few seconds than he had in days. We relive the incident on his behalf. We represent the words he had given us earlier, no longer knowing if they are ours or his.

My mother puts the watch on her wrist, admiring how well it matches her wedding ring. It was so very like him that he should spend his last quantum of mental energy organizing her gift, that he should exhaust his last physical effort in writing the card. At the first discussion of a birthday present for Mum, Ian's words were: 'This is going to kill her when he gives it to her,' and in a sense he was right. The shiny new Rotary marks the end of all those surprise presents and the beginning of Christmas-tree branches that will be forever empty.

We try, once he is asleep, to decipher what is written on the card. The scratchy black letters are poorly formed and indistinct, but it is clearly his writing and they are clearly his words. 'Cheap and Cheery, love and love, kisses and kisses, Bill.' This lavish timepiece was deliberately chosen to replace the previous watch, which, at $15.00, was not an extravagant gift by anyone's standards, certainly not by his. He had wanted to buy her something lasting and expensive at the time, but Mum picked the one she wanted, remarking that it might be cheap and cheery, but she liked it.

Her expensive new watch puts numbers on the artificial nature of being here and now. Nine, ten, eleven o'clock. He sleeps. We wait. I watch my family from some other place, looking down on our whole world with a profound sadness that oscillates from stoicism to gripping fear. The McKernans are just a bubble and there are bubbles upon bubbles upon bubbles. The world is getting away from me, expanding faster than we can imagine. Physicists call its boundaries an 'event horizon' and they liken the edge of the world to the surface of an ever-expanding bubble. Out there are other families, a multiverse of people dying, being born. Somewhere laughter and happiness exist on an event horizon that will never again see my world.

I sit and watch the hours pass me. This moment is just a moment and, here, it will soon be over. But, as any physicist would know, this moment will be happening over and over again for other observers in other galaxies, should such exist. Time is a fundamental property of the relationship between the universe and the observer. It was Einstein's math teacher, Hermann Minkowski, who allows us to think in the concept of space-time, coining the phrase to illustrate how the two are inexorably linked. *When* something happens depends on *where* you are. When the Canadian astronomer Ian Shelton saw a bright 'new star' in a distant galaxy, Large Magellanic Cloud, he was witnessing the energy released when a huge aging star blew apart, and he would say that supernova 1987A died on February 23, 1987. In fact, the star actually exploded in 165,000 BC, but it took until my father was fifty-four for its light to arrive here on Earth. New features of its death are still appearing, since stars die slowly in our perception of time. For inhabitants of Large Magellanic Cloud, if there are such beings, my father isn't dying yet. In fact, they'll have to wait 167,000 years for him to be born. And there is a place, not too far beyond Earth, where supernova 1987A and my father will die together.

Minkowski's work allowed us to see the universe as a 'block', rather than a sequence of events. Everything exists en masse yet we experience it sequentially, he said. The order depends on where in the universe we are, like watching a film strip, as the American philosopher and psychologist William James once said. At any single place in the universe (say, Saffron Walden, England, United Kingdom, Earth, the Milky Way, the Cosmos, as my daughter put it) the order of events does not change. My father was born before my mother and she was born before me. It's pretty well certain that his death will precede mine and the sequence here will always be the same. But while 55 Audley Road is seeing the end of this reel, the opening credits are only just playing on Mizar, the star at the bend of the Big Dipper's handle. Or, in the words of Albert Einstein, 'People like us, who believe in physics, know that the distinction between past, present and future is only a stubbornly persistent illusion.'

We are here and now and human. It is three thirty in the morning in a room six meters by five, coated in white paint and green Axminster and it's filled with little things that mean something. Over the fireplace hangs 'The Cricket Match', a print by Graham Clark that we bought for my father's sixtieth birthday. A Beleek basket-weave bowl, a wedding present from over forty years ago, sits on the shelf. Even by the dim light above my chair I can see my father's personality in the shelves of bedtime books: a cluster of Bill Brysons, a few political biographies and a generous selection of Wilbur Smiths. There is also a dust-free space on the second shelf down where a wrinkled old book with *A Time to Die* in decrepit lettering along its spine used to stand. It disappeared about a week ago and we have each of us—my mother, my brothers and I—sworn to one another that we have not moved it.

My father becomes fitful. His face flinches, his mouth grimaces. He gesticulates at something; tries to wave away imaginary flies, but his hands barely lift off the bed. Words blend together until they are no longer words but tones and notes of language; the sounds of a baby. He intermittently clenches his jaw and his skin is so stretched across his face that his lips no longer meet. His teeth and gums are permanently on display and the smile they portray is not like any smile he ever intended. It isn't a human smile of happiness and pleasure, more like the aggressive teeth-bearing of an ape.

'Are you in pain?' I ask, expecting no reply from the sleeping man. And, from nowhere, he replies with the strong voice of the person I have known all my life and with perfect clarity and coherence. He summons all the life that is left in him. And if I had only heard his voice I would have mistaken him, in that instant, for the man he was a year ago. He isn't Wullie or Mac or Bill. He is my dad, my father. The dynamism, urgency and drive that characterized his entire life are distilled in the words: 'God, Ruth—am I in pain. Get your mum.'

She doesn't want to come. She doesn't want to see him like that. But it is her duty to endure it painstakingly minute by minute, lifetime by lifetime, mopping his forehead with a wet flannel. I call the doctor's surgery for the emergency phone number and then I wake the poor man on duty. On his advice, I double the strength of the morphine in his syringe driver. My hands are shaking as I make up the solution and transfer it into the syringe but I can do it. It is a small task, the like of which I have done as a biochemist at work many times before. I can do it, but quickly, quickly, now!

More morphine will relieve the pain. There isn't a pain that can't be relieved by more morphine. But what if it doesn't? What do I do? The glass ampoule quivers in my hand. I can't judge the strength needed to break off the top. Will I crush the glass, spill the precious solution on the carpet? It breaks cleanly. How much is there? Certainly enough to double the dose for a day or so. What if I were to put it all in? Enough? Not enough? I calculate. I don't know. I don't know.

Thankfully, the higher dose of morphine helps a bit, but he is still grimacing and restless. Dr. Clayton-Payne from Gold Street surgery arrives, injects more and asks if the pain is subsiding. My father's vague nod is taken as agreement, and his deeper and deeper breaths drop him into a deeper and deeper sleep. How many left? How much time?

The number of breaths an animal takes in its lifetime is a surprisingly consistent number. A mouse lives for two years and its heart beats about 500 times a minute, while the elephant lives much longer and has a pulse of just 28. But the number that sticks in my mind is 1.5 billion. It doesn't matter what size an animal is, its lifetime lasts for about 1.5 billion heartbeats. Actually, the benefits of modern medicine and civilization have pushed the human limit a bit further. Even so, what the latest research from the Santa Fe Institute says to me (other than don't waste those heartbeats by using them up jogging) is that my father has been short-changed on a scale of millions.

According to the theoretical physicist Geoffrey West there is a precise relationship between the size of an animal and many of its functions. Normally, metabolic rate is proportional to the animal's mass raised to the three-quarters power and this holds true from amoeba to antelopes. Biology is being reduced to mathematics; eventually a formula might quantify evolution itself. Something sets the upper limit of a human lifetime at a little over a hundred years and the same equations work for all life on Earth.

'It won't be long,' the duty doctor says. Maybe ten minutes, maybe ten hours, but not days.

Billy's last molecule of ATP is spent. Without cellular energy, sleep gives way to coma. My father's eyelids drop but never fully close and we can see the yellow of the whites of them. His breathing is measured, rhythmical and very loud. Not that this is completely out of place. The timing of his breathing is different, though. His respiratory cycle has a meter quite unlike that of a sleeping man. It is more reminiscent of the mechanics of the ventilator than a human pair of lungs. At close to six his breathing becomes shallower, as folklore says it would, and we watch the lightness of it evaporate like a cold nitrogen cloud. His body pauses, draws a few more shallow breaths and finally stops.

In that instant the transformation from life to death is made and the experience of watching a man die is never to be forgotten. Despite days, weeks and months of anticipation, there is no preparation for the shock that separates that last breath from the unfulfilled expectation of the next. You think it is a small step but, no, it is a chasm that can never be bridged. And it made no difference that there was plenty of warning. I believed I was well prepared but the knowledge that my father would die was less than a beginning, less than a seed of the real experience. I thought I had started saving for this moment a year ago, kept up the installments for months, and that somehow the final payment would be quite painless. How wrong can you be?

I prayed to have the last decade, year, minute over again, to expand time out like the universe. Even the parts that seemed slow, as I lived them, have snapped back still faster, collapsing Billy's life into a singularity of no measurable dimensions. There is no way to measure it, or, put much better by W. H. Auden:

Clocks cannot tell our time of day
For what event to pray,

Because we have no time, because
We have no time until
We know what time we fill,
Why time is other than time was.

For me, my father no longer exists, in any physical sense of the word, from this moment. It is a few years since his CLL was diagnosed. More poignantly, it is almost nine months since he contracted septicemia and his body started shrinking. The deconstruction of life was a steady process, a reversal of fetal development. Most cell divisions occur in the first three months as life takes form. Most shrinkage occurs in the last three months when life dissolves. Billy's weight stabilized for a few weeks when he came out of the hospital but he never really put any back on—he just lost it at different rates, accelerating toward the end. He went from over two-hundred thirty-eight pounds to probably less than eighty-four in exactly the same time as it takes a fertilized egg to turn into a self-sustaining independent being.

The pattern is the same for a cell, a body, perhaps even, on a much grander scale, for a universe; an explosion of birth, a functional, stable, well-orchestrated life and then accelerated decline to less than nothing. Ultimately, my father withdrew from his body, gradually detaching himself from the casing that held him until, with his last light breath, he collapsed in on himself.

In death, his body is more like art than reality: an object, mineral and inanimate, with an outer covering of fine, dull gold. I look away from it and into the room. The well-worn, maroon leather slippers that once warmed his feet loll untidily at the foot of the bed. The spectacles he wore perched on his nose and took on and off with irritating frequency rest on the bedside table. Next to them sits the child's beaker that nestled between his tired, trembling lips to deliver those last few sips of Highland Spring. All these things are more him, or part of him, than the leftovers on the bed.

We go downstairs, unable to spend any more time in the same

room as his empty carapace. We leave the syringe driver feeding morphine into the top of its arm, as if to ensure that it can feel no more pain.

It is over.

When Francesca comes home from school, I sit her down next to me on the sofa and put my arm round her. 'You know that Granddad hasn't been well for a long time,' I say as softly and gently as I can manage. 'Well . . . he died this morning.'

She runs into the playroom and grabs her favorite puppet, Lamb Chop, curls up on her beanbag in a fetal position and sobs. I try to cuddle her but she pushes me away. No more than two minutes later she is back, climbing onto my lap, still clutching the animal to her chest.

'I didn't understand before,' she says, in a more composed manner than I would have thought possible. 'And now I do.'

It is really over. He has slipped into the universe.

But where exactly in the universe? As a small child, no older than my daughter, I would look up at clouds and wonder which one contained heaven. We thought the universe was smaller then, before we knew about the Big Bang or the ten dimensions that string theory predicts. It would have been even smaller when Walter de la Mare wrote that 'the universe is not composed of dead matter but is . . . a living presence.' Now scientists propose that something out there is living, but we see no Botticelli heaven, only bacteria.

Starlight is dimmed by a perpetual, hazy cloud of dust. Sir Fred Hoyle, the British astronomer, and his Sri Lankan colleague Chandra Wickramasinghe argue that those dust grains are bacteria. They are the same size as bacteria, have the same spectral, or light-scattering, properties and the physicists calculate that since the universe was formed twelve to fourteen billion years ago, enough bacteria could easily have been generated to account for all the dust there is.

Looking out into the universe we see the cosmic dust, yet the major component of the universe is something that we can't see, yet we know exists: dark matter. There is ten times more dark matter than all the stars, planets and galaxies combined. Astronomers know that the mysterious form exists because they can detect its effects. It has gravity. It changes the behavior of planets. It bends light. We know it's there but we can't measure it. It emits no light, no radiation, no observable energy of any kind. What it is, or what it will become, remains one of the great outstanding questions of astronomy.

My father becomes the space that fills the house, huge and dominating. We move around his massive presence but can no longer see or hear him. The whole world is proof of his existence but there is no device on Earth that can show us where or how he is. He used to be ordinary matter, the stuff of stardust and clouds and gas, with some heavier elements—metals, halogens—thrown in. What is he now? What is a star when it is gone? What else is there?

He has become something I don't understand. He is no longer a living presence; he is cold, dark matter.

12

Distance

Maybe all this
Is happening in some lab?
Under one lamp by day
And billions by night?

Maybe we're experimental generations?
Poured from one vial to the next,
Shaken in test tubes,
Not scrutinised by eyes alone,
Each of us separately
Plucked up by tweezers in the end?
   From 'Maybe all this', *Poems*,
by Wislawa Szymborska

SCIENTISTS DON'T REALLY KNOW MUCH, IN THE SENSE OF BEING absolutely sure, being 100 percent positive. We are 95 percent kind of people. We have hypotheses and when most of the data fit we believe we can be fairly conclusive. We get ever closer to a fact but there is always that little bit of room left for doubt, just in case someone comes along with a better idea. We have our own phrases to cover such thinking. The scientific literature is littered with sentences that start 'There is strong justification for . . .' or 'All the current evidence supports . . .' or 'In all probability . . .'. This time is an exception. For once, humbly, I know.

Knowing is what makes us human. We don't exist on mere instinct or trained associations; we have evolved further than that. The

wonder of researching the brain, marveling at how we—our person-
alities and minds—are created, brings with it the unavoidable knowledge that each of us is mortal and each of us will die. As a species,
though, we continue to evolve. The Cambridge Professor of Behavioral Neuroscience, Barry Keverne, maintains that 'in humans, evolution has favored executive function over emotion'. Does this mean
that perhaps one day we will be smart enough to care less about what
pains us? Will we grow out of grief?

Can it be true? What about the most analytical of humans—
Einstein, for example? Was he clever enough to lack emotion? His
brain was retained after his death and it has been investigated in
many ways. The most recent study was published in the medical
journal *The Lancet* in 1999 and concluded that the genius had an
unusual pattern of grooves in his expanded parietal lobe, which could
arguably account for his mathematical proficiency. But a man who
could say 'How on Earth are you ever going to explain in terms of
chemistry and physics so important a biological phenomenon as first
love?' seems not to be trading executive function for emotion.

I asked Professor Keverne what he meant. He explained
that he doesn't expect humanity to evolve into automatons. Emotion won't go away. Rather, we are fortunate in being able
to anticipate rather than just react; to eat before we are hungry, to
talk our way out of trouble and to understand why we are unhappy.

Scientists never say never. We mean never in the same way that
Paul McCartney meant never. When he was released from court after being charged with possession of cannabis, a reporter asked if he
had a message for the youth of today (well, the youth of the time:
most, like him, would be grandparents now). He looked directly
into the camera, gave a subtle wink and declared: 'I'll never do it
again.' That's what never meant. I'll never take biscuits without asking. I'll never be late home again. I'll never turn up for another meeting unprepared. Never was little more than a declaration of intent.

Not any longer. Never is an eternity and, as Woody Allen said,

'Eternity is very long, especially towards the end.' Never is time and distance that cannot be revisited; the stegosaurus, the Holy Roman Empire, the ground floor of the twin towers, my twenty-first birthday. Non-negotiable. Gone. Over.

I know I'll never see my father again.

There have been times when the absoluteness of that knowledge tormented me. How wonderful it would be to be able to know something different, to believe in some other, better option. I would drive to work on a crisp winter day, think of him and feel the tears spill onto my cheeks. Each spring, my resentment for daffodils grows stronger; a reminder of the day I angrily kicked off their joyful heads and ground their insolent trumpets into the soil for the crime of appearing out of the ground so cheerfully just weeks after he died.

The feeling was most acute the first time I returned to Sarasota. I listened to his voice greeting callers on the answering machine. 'Well, it's best not to change it,' my mother said. A man's voice is always a good deterrent if you're a woman living alone. I stared at the empty chair where I had so carefully taken that last photograph. How I fantasized about what I would give to see him sitting there, in the flesh: to be able to talk to him, to watch his mouth move and hear his soft Scottish voice say something more than: 'You have reached the McKernans in Florida. We can't come to the phone right now, so please leave us a message.' My right arm would be the classic reply—I would give my right arm to have him back. Except that my right arm would be a lot to give just to postpone the inevitable. Maybe a hand or a finger? What weight of flesh for a day?

I could tell Billy the essentials: Mum's new bungalow is perfect for her; his sons are doing just fine running Molecular Products without him—not as he would have done it, naturally, but fine nevertheless. I could tell him the story of my promotion to become head of the Research Center; how I hadn't taken the succession plan seriously. I had dismissed my boss's plan to make me his heir as all theory until two years were condensed into two months and swiftly became 'next Monday'. Dad would have been proud. He wouldn't have said so, but he would have found a way of showing it. I could tell him about the progress that has been made on drugs for Alzheimer's disease. Not that he'll ever need them, but he'd be interested just the same. I wish I could have discussed with him my decision to change companies, although perhaps he would have encouraged our move to Kent just for the opportunity to play golf with his old chums when he visited. I would tell him that the house in Sarasota is as lovely as ever; that the palm trees reach as high as the roof, that there were two red cardinals at the bird feeder today. We identified them from the bird book he gave Francesca the last time we were there together. I would say that my children have grown to love the place as much as he did. I'm sure I could get all that in for the price of a pinkie.

When I look at that last photograph now, it tells another story. It used to remind me of the day my father died, when I took it into the bathroom and wallowed in painfully hot water, just to feel something else. I put the picture on the shelf beside the bath and stared at it, drained it of all the comfort there was to be had. Now it sits in a silver frame beside my bed and its presence marks the passage of time. I see my father and daughter smiling at each other as though happiness is perpetually theirs for the taking.

Things look different from a distance.

Francesca's podgy cheeks and girlish innocence have been swapped for hormones and attitude. Adam can tie the laces of his soccer cleats himself and he no longer even notices the scary fancy-dress costumes of the Sarasota Mall when we are there again at Halloween. The mental picture of my father has changed too. He hasn't aged at all, nor shall he. I often think of him looking younger than he was when he died, although for many months I could see nothing but his corpse.

Those images of illness, apoptosis and pain have faded, shifted by the Doppler effect of experience. Remember the Doppler effect?

It is the way sound can change with movement, named after the Austrian physicist Christian Doppler who discovered it in the mid-1800s. When an ambulance is moving toward you the sound of its siren shifts to a higher frequency; as it moves away, it becomes a lower tone. The same thing happens with light. If a light source moves toward you it shifts toward blue; as it moves away, it becomes more red. The same was true of my father's death. I could see it coming for a long time; a blue ball of anxiety, pain and fear. For the snapshot of time over which it happened it occurred in black and white and every other super-real color imaginable. Now, as my father's life pales into the cosmos, it is surrounded by the warm red glow he leaves behind.

Yes, things look different from a distance.

While Billy lay dying, I sought solace in science. But being a scientist spared me nothing; it provided no shield against fate, no defense against grief. Knowing how emotion affects memory could not keep out the vision of those last few breaths. Knowing how genes and development together mold personality could not make me love my father more. No matter what we understand of the human mind, it doesn't change what we are.

The more we learn, the smaller we appear. From our planet, the rings around Saturn look like a single entity, even though the latest pictures from Voyager 2 show that there are really seven of them. From here the punctuated loops blend into one another. From Neptune, Earth is barely detectable; the smallest visible unit in the Milky Way. It prompted Carl Sagan's immortal observation: 'We live on a blue dot.' From that blue dot radiates my family's entire past and future. On the universe's scale we are no more than butterflies whose lives last just one day. Everything that we are, everything we believe we might be, comes from a world no bigger than a pixel; no bigger than a pixel I might see on a brain scan; no bigger than a pixel generated by repetitive learning or revisited memories—the pixel swollen by my father's frequent kisses.

As Sir William Bragg said, 'The important thing in science is not so much to obtain new facts as to discover new ways of thinking about them.' Butterflies we may be, but we have big brains with huge cortices and cognitive power beyond compare in the world we have explored thus far. We may be no more than genes, self-assembled into lifetimes, that become history; overlapping, ever-growing, concentric rings. Out there, somewhere, Billy's ever-expanding halo will fade into mine and then Francesca's and so on, merging into one warm, glistening, pride-filled orb.

Index

Index9878